Legal Disclaimer

Copyright 2017- All rights reserved.

In no way is it legal to reproduce, duplicate, or transmit any part of this document in either electronic means or in printed format. Recording of this publication is strictly prohibited and any storage of this document is not allowed unless with written consent by the publisher. All rights reserved.

The information provided herein is stated to be truthful and consistent, in that any liability, in terms of inattention or otherwise, by any usage or abuse of any policies, processes, or directions contained within is the sole and utter responsibility of the recipient. Under no circumstances will any legal responsibility or blame be held against the publisher for any reparation, damages, or monetary loss due to the information herein, either directly or indirectly.

Respective author(s) own all copyrights not held by the publisher.

Legal Notice:

This book is copyright protected. This is only intended for personal use. You may not amend, distribute, sell, use, quote or paraphrase any part or the content within the book without consent of the author or copyright owner. Legal action will be pursued if the previous are ever breached.

Disclaimer:

Please note the information contained within this book is for educational and entertainment purposes only. Every attempt has been made to provide accurate, up to date and reliable complete information. Under no circumstances does this book express or imply any warranties of any kind. Readers must acknowledge that the author is not engaged in the rendering of any form of legal, financial, medical or professional advice.

By reading this document, the reader agrees that under no circumstances can we be held responsible for any losses, direct or indirect, which are incurred as a result of the use of information contained within this document, including, but not limited to- errors, omissions, or inaccuracies.

Everything Electrical: How To Test Circuits Like A Pro: Part 1

Preface:

Have you ever studied electricity in high school, a college class or maybe a trade school and felt like the teacher didn't tell you everything you needed to know? Or that the theory just didn't give you anything useful to use out on the field? That they didn't prepare you for those uncommon or intermittent electrical issues that leave you feeling like you don't have a plan of attack. Well either way GREAT! You are not alone. I myself read at least 10 full textbooks on electrical, electronics, industrial electrical and automotive electricity, that by the way were not very cheap averaging in cost around 150$ each. But these books still left me feeling like they failed in many aspects for learning real world electrical tips and tricks.

This book was written to educate in a simpler way for everyone to understand, beginners and veteran technicians alike. There is no reason to complicate things with big words that usually are left unexplained by other books and make it even harder to understand with bad examples. This book is priced low but because I feel that everyone should know at least the basics. I will include many examples of each topic I discuss for better understanding. Because of my approach to certain topics, I recommend that you read the book front to back even if you feel you've already read too much theory of electricity. My goal is to make you "the electrical guy" that will fearlessly tackle any job. If this book series "Everything Electrical" does not teach you everything you wanted to know, I guarantee that it will at least be a very powerful supplement to your learning of electrical testing at a low price.

This book is part of a series on how to use your meter like a professional electrician and/or technician. To take full advantage of this book you must already know how to use the settings on your multimeter in at least a very basic way before reading this book. If you do not already know how to use a meter I will try to review the key concepts of the most important meter settings but I strongly recommend that you read my other book first "Everything Electrical: How To Use All The Functions Of Your Multimeter" before starting this book.

Table Of Contents:

Ch.1: Important Things To Remember Before Starting..*(1-18)*

Ch.2: Voltage Testing Like a Pro........................*(19-24)*

Ch.3: Open Circuit Testing................................*(25-34)*

Ch.4: Resistance Testing Using The Voltmeter........*(35-43)*

Ch.5: Miscellaneous Tips...................................*(44-59)*

Conclusion: (Summary & Ending Words)............... *(60)*

Ch:1 Important Things To Remember Before Starting

Before we can begin our testing, we need to make sure that you understand at least the basics of electricity. You may have already learned this from other experiences, but for the sake of learning and fully understanding all of the testing methods in this book, I will do a quick review of the basics.

Electrical Circuits:

Let us begin the book by looking at a sample of an electrical circuit. A basic **electrical circuit** is the combination of a power source, electrical wiring, a fuse, a switch and an electrical device, connected in a way that makes the device work. In order for any electrical device to work it must first be connected to a complete electrical circuit.

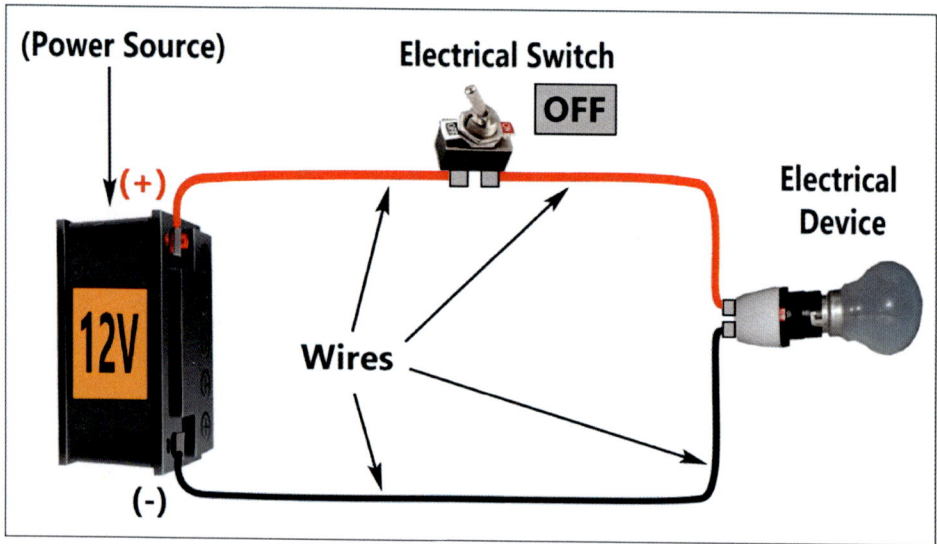

(In this illustration, you see the main parts of a basic electric circuit. A circuit has a power source, in this case our battery. A switch for control of when the circuit is ON or OFF. The electrical device that is being powered on, this case our light bulb. And also, a set of wires with one going from the positive side of the power source to the electrical device and the other from the electrical device back to the negative side of the power source. All these parts put together are what make up a simple circuit.)

Note: The two wires going to the electrical device from the main power source are called the **power wire** and the **negative wire,** respectively. The one supplying the voltage is known as the power wire, and the other low voltage wire is called the negative or return wire. Other names for the power wire include the **feed, hot** or **live** wire. Other names for the negative wire are the **return, earth, ground** or **voltage low** wire. Keep these in mind for the future.

Now let's take a look at the minor yet still very important parts of a circuit that I haven't mentioned yet.

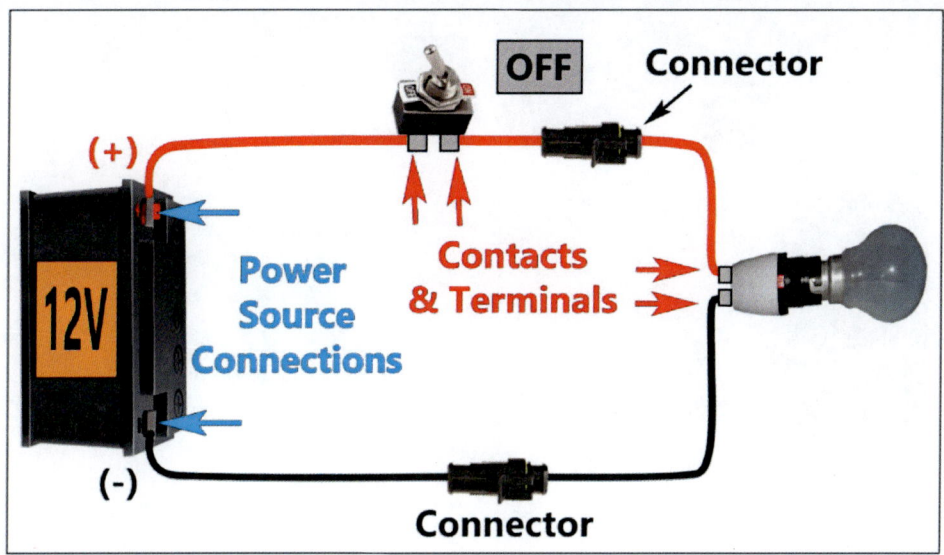

(In this illustration, we see the minor parts inside of a circuit. These include the connections to the power source, in this case our two battery clamps. Also included are two connectors in between the wires of the circuit and every other contact and terminal in the circuit that fits together to make a secure electrical connection.)

Always Check Connections: Because they are minor parts they are very often overlooked when troubleshooting an electrical issue. Remember that a bad contact or loose electrical connection will result in the circuit not working right or not even working at all. Always check for loose connectors, terminals, electric clamps and wires first! If you forget about these minor parts of a circuit you might struggle to find the cause to your electrical problem which could have been an easy fix.

The easiest way to check for bad connections is to simply wiggle them to see if there is any looseness that could be the cause of the electrical problem. This basic test is known as the **Wiggle Test** and should never be forgotten when diagnosing an electric problem. Very often, a problem is easily fixed by someone without experience, simply by checking connections. The job is finished before ever taking out a multimeter, just by wiggling a few connections to check for looseness and making sure that all the connections of the circuit are on tight. DONT forget the little things!

Now let us look at some additional parts of circuits you will commonly come to see...

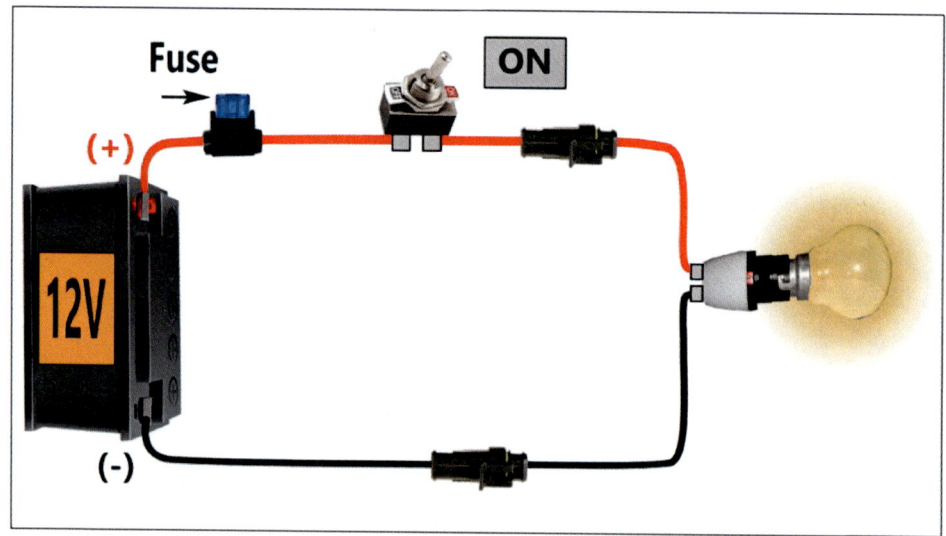

(This picture shows our circuit with a fuse added. The fuse is designed to blow whenever there is an electrical short or a surge in the circuit. This stops all electricity from flowing in order to protect the circuit from damaging itself due to the short that exists.)

Almost always you will have some kind of circuit protection device installed in a circuit to protect against shorts or surges. You will very likely see a circuit protection device during testing, whether it's in the form of a **fuse** or a **circuit breaker**.

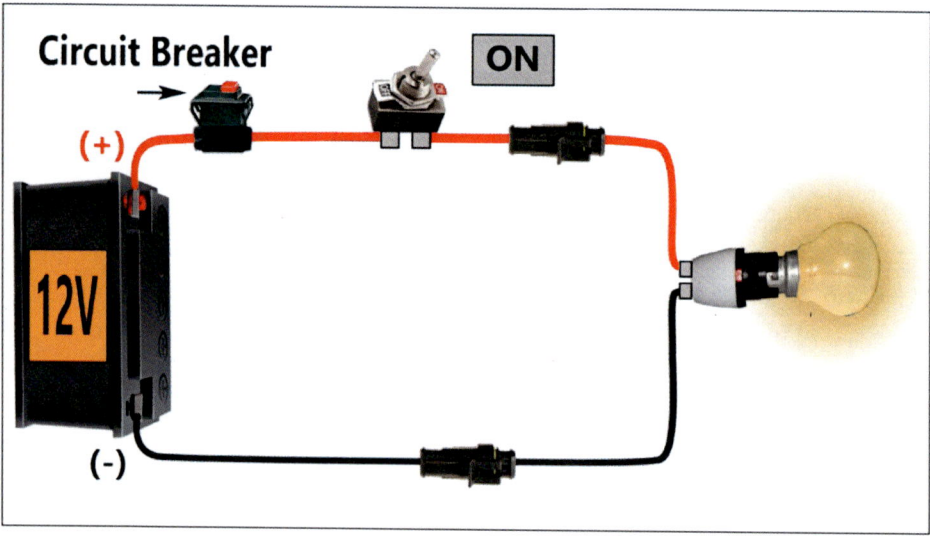

(This picture shows our circuit with a Circuit Breaker instead of a fuse. This is just another kind of circuit protection device.)

Just like with a bad connection, a circuit protection device installed in a circuit has the potential to become a problem. For this reason, we must also learn how to test them properly. We will explore how to check the circuit protection device along with other parts of a circuit later on in this chapter. For now, lets continue on with more circuit variations that you will commonly find.

Variations of the Basic Circuit:

Next are examples of common variations to the basic circuit.

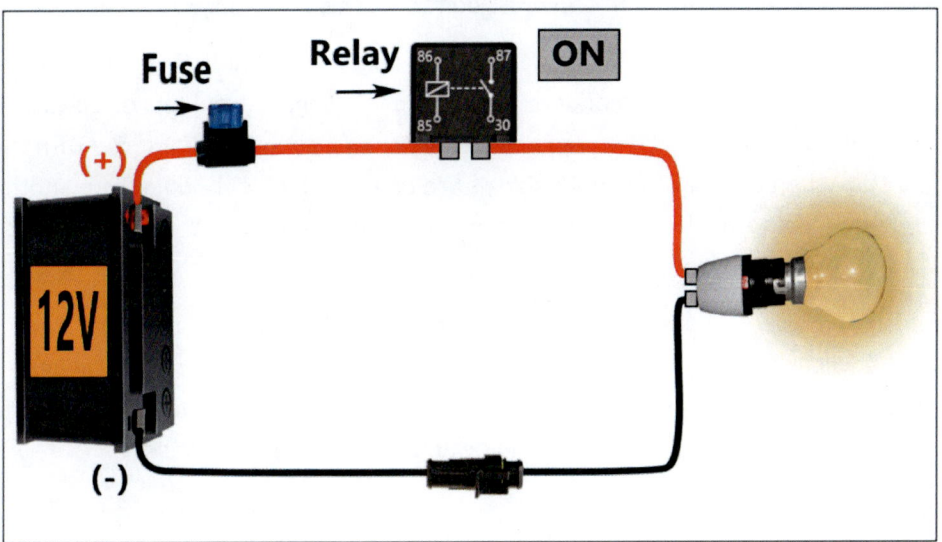

(This picture shows our fused circuit with a relay added in place of our electric switch. The relay works as a kind of switch to turn the light bulb on. It turns on the light bulb only when a second circuit that is involved with this circuit is turned on.)

A relay is not so easy to explain or test easily because it always includes testing at least two or more circuits that are involved with each other. For this reason, I have written another complete book on these types of complex circuits and how to test them. If interested, please check out my other book "Everything Electrical: How to Test Relays And Involved Circuits".

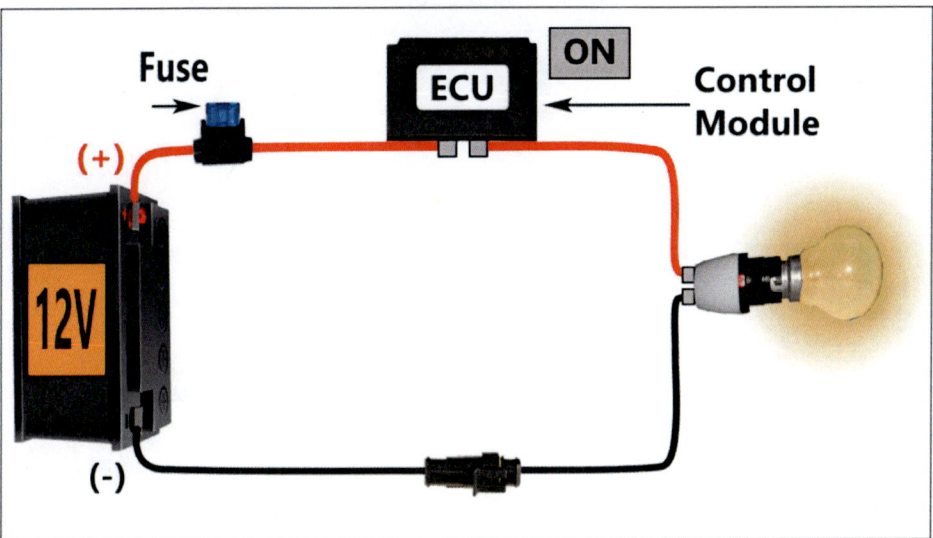

(This is another variation of our common circuit. This version of our circuit involves the switch being replaced by a control module. The control module acts as a smart switch for the circuit which will only turn the light bulb on when the module gets the signal to do so from another circuit or sensor that is involved.)

Many circuits can have a fuse, a circuit breaker, a control module and/or a relay all in one circuit. These parts are all to be considered when diagnosing a problem. Whatever the circuit has in it can be a potential problem and must be thoroughly checked.

Note: The module in the previous example is responsible for turning on the light bulb of the circuit. Circuit computers or Commanders, Programmable Logic Controllers or Control Modules all do the same thing. They replace the switch in a circuit so that it can turn the circuit ON only when the required input signals are received by the module from another circuit or sensor that is involved. Think of a control module as a "smart" switch that turns on a circuit only when it is signaled by another circuit's outputs.

The purpose of these examples were to simply introduce you to what a real-life circuit may look like. We will explore and test the more complex circuitry in another more advanced book of this "How to Test Like a Pro" series. For now, let us first continue on with the review lesson of the basics by defining the three major units measured in electricity.

Voltage:

First is the unit known as voltage, which is the amount of electrical pressure inside of an electric circuit. The **electrical pressure** is needed in order to make electricity travel throughout the circuit. Without voltage in a circuit, no electricity will be able to flow. Voltage can be compared to water pressure, provided by a water pump, that makes the water flow inside of a water pipe system.

By checking for the voltage available at various points in a circuit we can gather a lot of information about the condition that the circuit is currently in.

The tool required for measuring voltage is known as the **Voltmeter**, or the multimeter set to the volt setting. The following are some examples of a voltmeter measuring voltage.

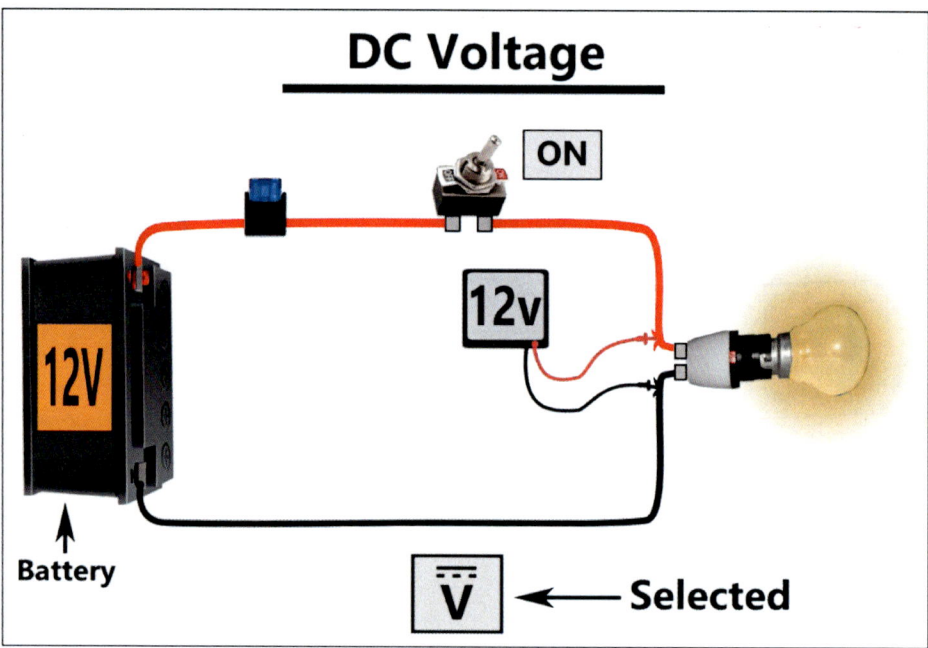

(This example shows a voltage reading being taken at the electrical device of the circuit. The meter reads 12 volts DC available up to the electrical device. The multimeter is set to read DC volts because the power source in this circuit is a DC power source.)

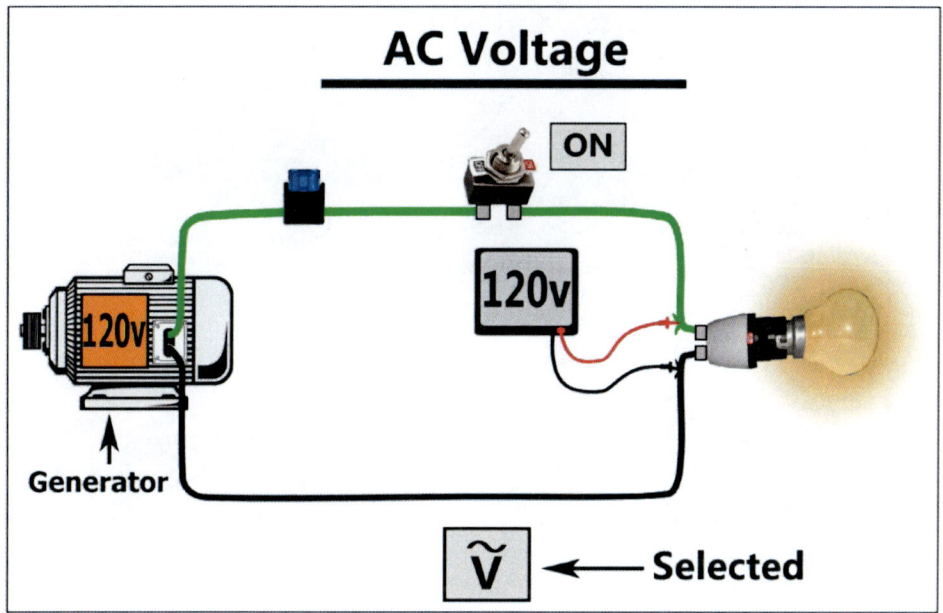

(This example shows another circuit's voltage being taken at the electrical device. The meter reads 120 volts AC available up to the electrical device in this circuit. The multimeter in this example is set to read AC volts this time because the power source in the circuit is an AC power source.)

As seen in the previous images there are two types of voltages that exist. There is DC voltage and AC voltage. Regardless of what type of voltage it is, the circuit testing is still the same. The only difference is in the meter setting that you use depending on whether the circuit is running off of AC voltage or DC voltage.

The **AC voltage** setting on your meter will be used when measuring voltages from circuits powered by an AC power source. This commonly includes household power outlets, household wiring, Industrial wiring and/or practically any other electric circuit powered by an AC generator.

The **DC voltage** setting will be used when measuring voltages from circuits powered by a DC power source. This includes many automotive, motorcycle, aviation, some industrial electric circuits or any other circuit that is powered through a battery, capacitor, a solar panel or a DC generator.

Now let us take a look at a few more examples of what the voltmeter is commonly used for.

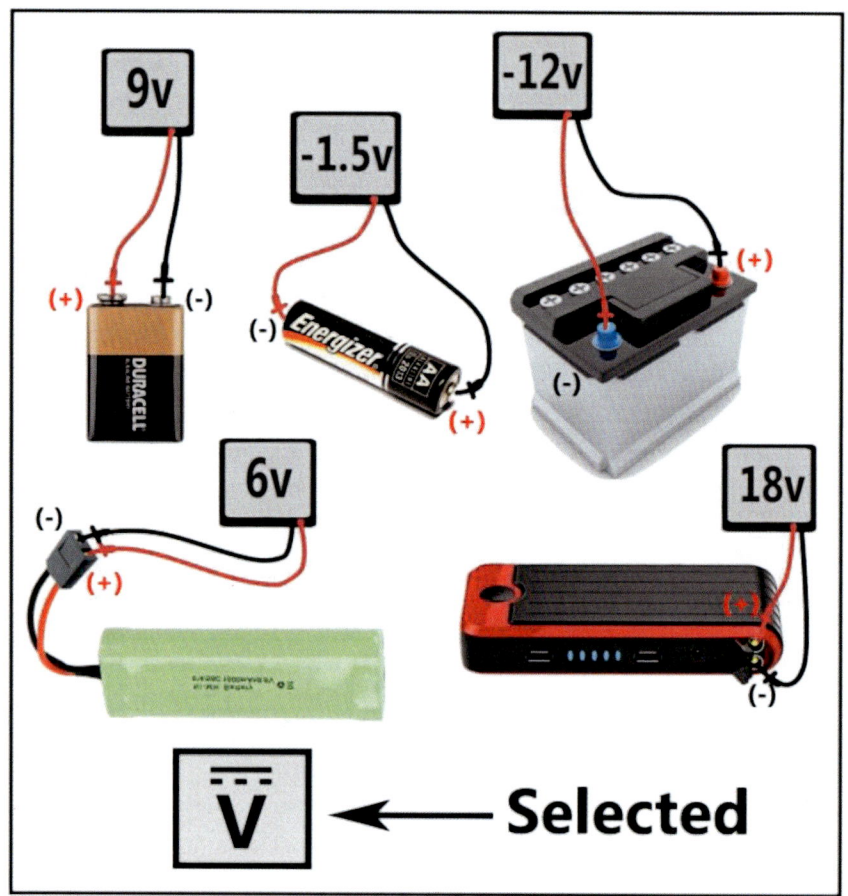

(This image shows examples of the voltmeter being used to perform bench testing on various batteries. The batteries being tested should contain the voltage that they are labeled to have.)

Notice the reading on the 12-volt battery as being negative. This not a mistake, it just means you have your test probes connected backwards. No damage will be done to the voltmeter if you mix the test probes up, it will just display the voltage of the battery in a negative value.

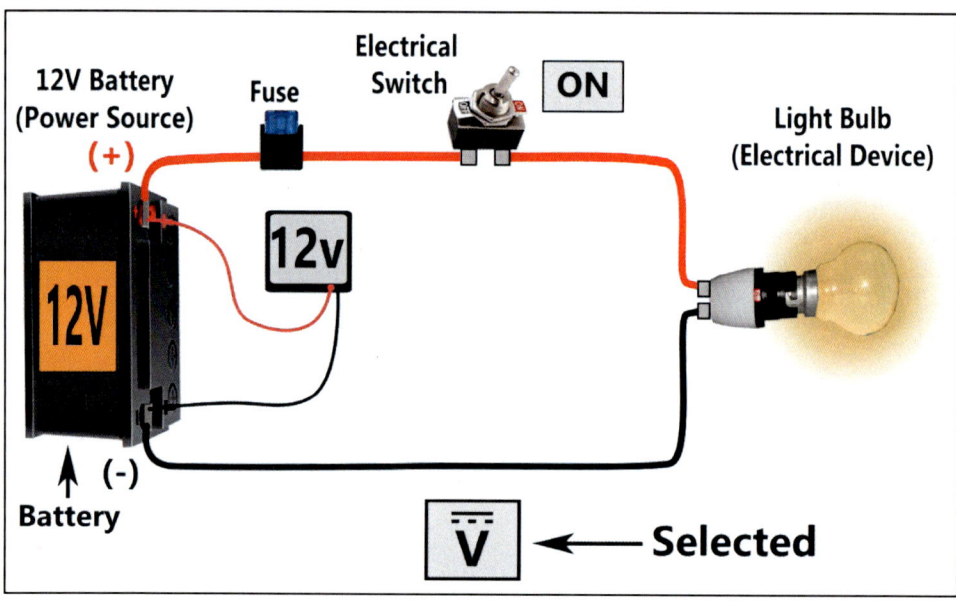

(This illustration shows a voltmeter being used to check the power source's voltage while it is connected to a circuit. In this circuit the power source is a 12 volt battery. The voltage is being checked while the circuit is ON and working.)

By testing the power source while it is powering a circuit, it allows you to see how well the battery is performing when it is actually being worked and drained. If the battery voltage is ok during a bench test, but goes very low in voltage when it is installed for powering a circuit then the battery is most likely bad and needs replacement. This problem commonly happens in a battery that can't hold a charge anymore.

REMINDER: In order to get any kind of voltage reading, you will need to make metal-to-metal contact with the conductive parts of a circuit using your meter's test probes. This can be done by probing at terminals, clamps, connectors or by piercing the wires of the circuit carefully to make contact. If you look closely at the first two images in the beginning of this voltage lesson, you can see that the meter's probe tips have pierced through the wire to make contact with the metal part of the wire. The best way to actually do this without damaging the wire as much is to use a special add-on tool for your multimeter called **wire piercing probes**.

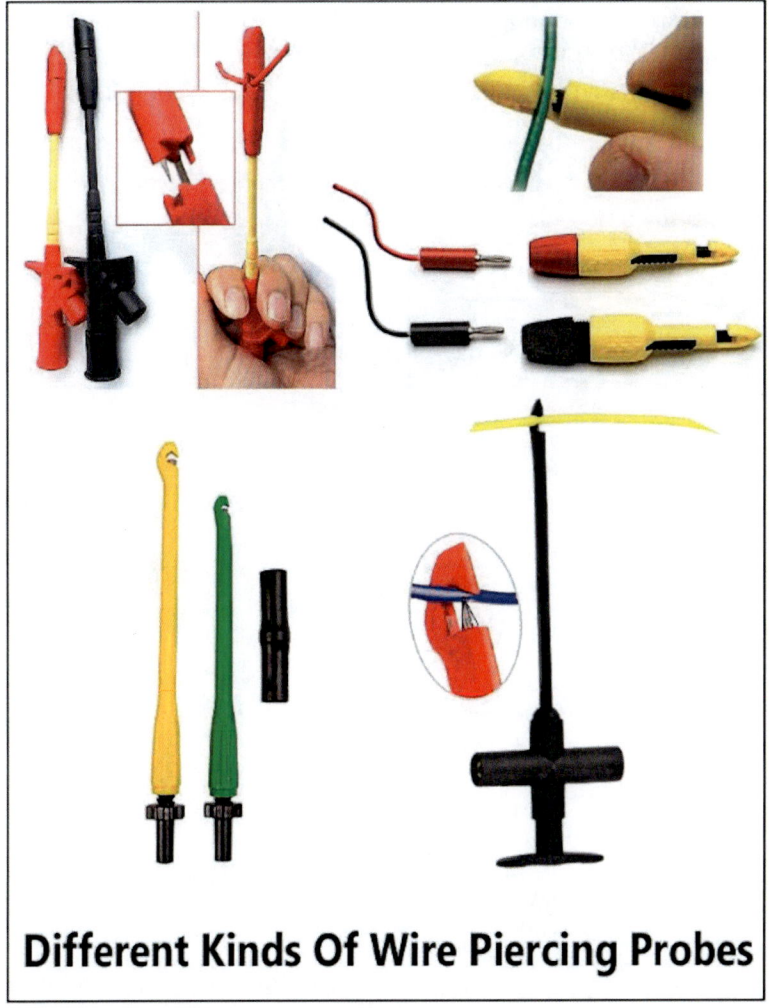

Different Kinds Of Wire Piercing Probes

(This illustration shows different examples of how the piercing probes connect in order to get a stable voltage measurement from a wire. The top image shows the needle tip of the test probe, piercing the wire to make contact. The other images show other kinds of wire piercing probes that are available for your meter.)

The biggest benefit from using these types of probes is that it makes the job easier and doesn't require you to hold the probe in place to make sure you get a good connection and measurement. Just remember that once you are done testing using these piercing probes, to always put a piece of electrician's tape over the hole you made during testing to prevent corrosion of the wire from exposure to the environment.

Amperage:

Next on the list of major electrical units is amperage or **Amps**, which is the amount of electricity that is flowing inside of a circuit. The textbook definition for amps will say it is the amount of electrons flowing per second per inch through a wire and although this is true, let us just define it simply as the amount of electricity that is flowing inside of a circuit. Amperage in an electric circuit can be compared to the amount of water flowing inside of a water pipe circuit. Do not confuse amperage with voltage. Although they are related and similar they are not the same thing. Remember that voltage is electrical pressure and that amperage is the actual amount of electricity flowing throughout the circuit because of the electrical pressure. Amperage cannot flow without voltage to push it throughout the circuit. Just like how the water inside of a water pipe circuit will not flow unless there is water pressure being provided from the water pump.

The tool used for measuring amps is the ammeter, or the multimeter set to the correct amp setting. By checking a circuit's amperage, this allows us to see the amount of electricity flowing inside of the circuit. Let take a look at some examples of the ammeter measuring amperage.

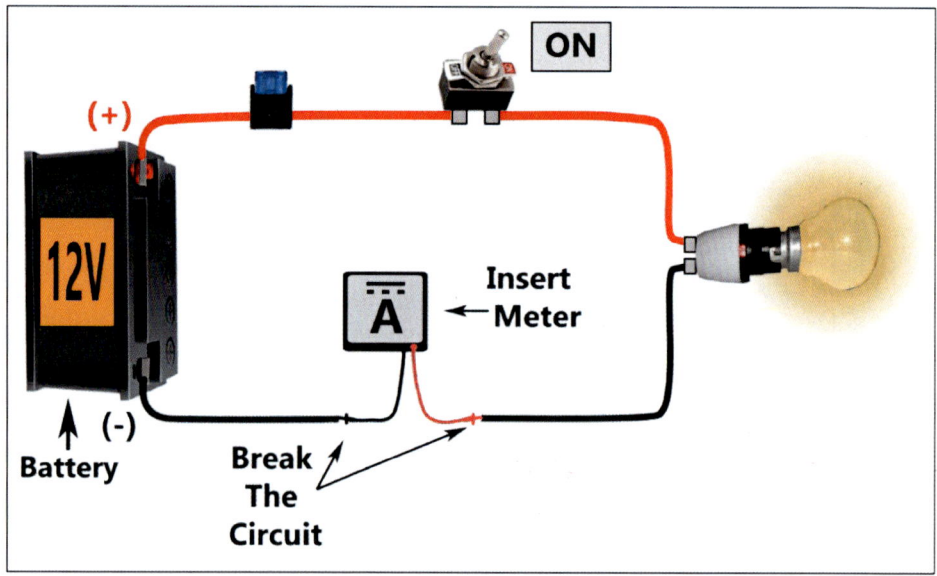

(This example shows how an ammeter is installed into a circuit to read **amps**. In order for the meter to read how much electricity is flowing you have to make sure it becomes part of either the negative or positive side of the circuit so that the electricity can flow through it and be measured. Think of the ammeter as a set of jumper wires with a gauge to read the amount of flow of electricity. The ammeter is, in a sense, a flow meter. In this example we decided to install the ammeter on the negative side of the circuit.)

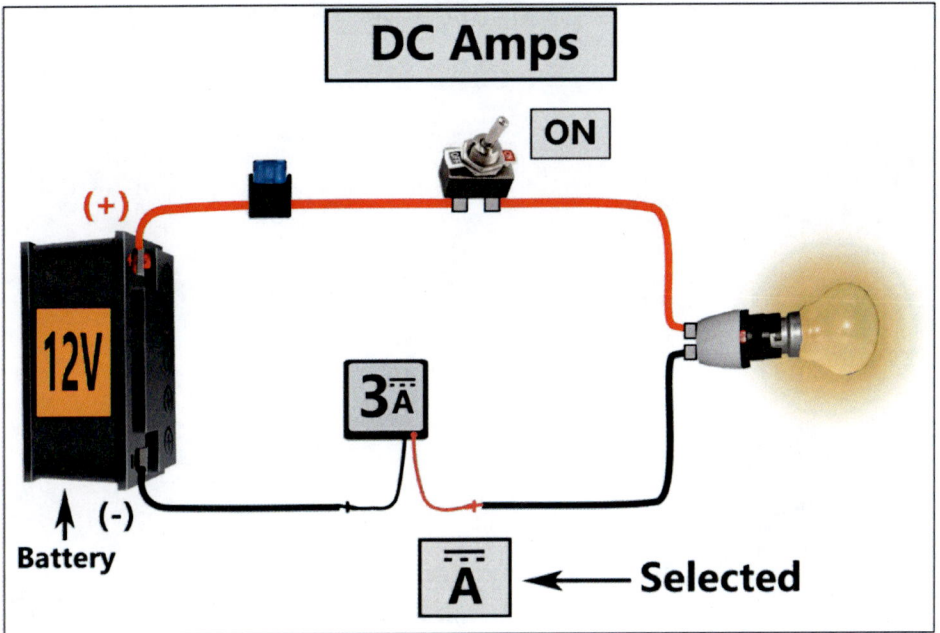

(This example illustrates the ammeter reading 3 amps of current flow, after it has been installed into a DC circuit and set to read DC amps. Whenever you have a DC power source such as a battery you will need to set your meter to read DC amps.)

Whenever installing the ammeter, try to break the circuit in a place that is easiest to access. One easy way to install an ammeter is to disconnect the circuit at a connector or terminal and put the ammeter in between the two terminals. This will ensure that the electricity flows through the meter and is measured.

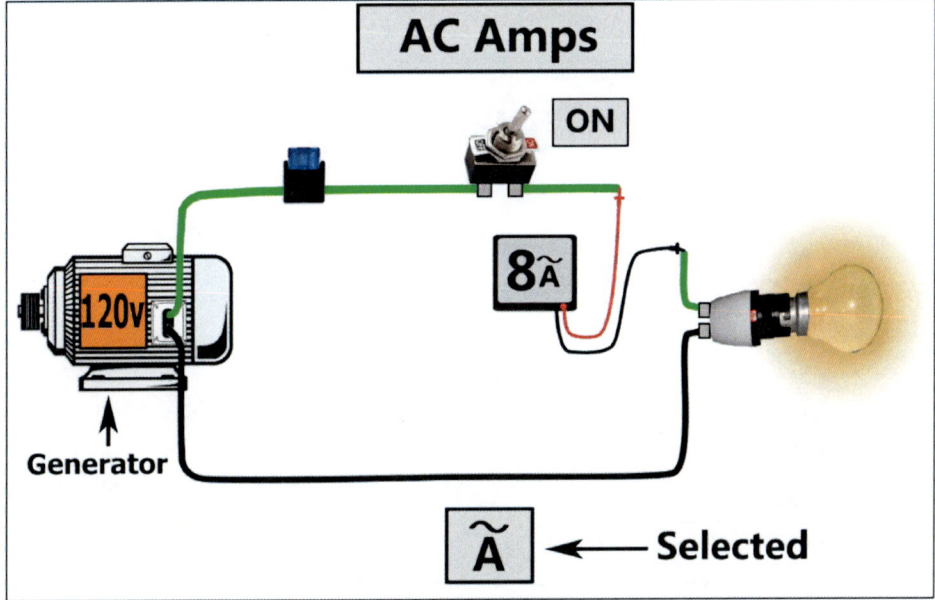

(In this example, the ammeter was set to read AC amps and installed into an AC circuit on the power side. Whenever you have an AC power source such as an AC generator you will need to set your meter to read AC amps. Try to break the circuit and install your ammeter in a place that is easy to access to make for a quick installation and removal. The same reading will be seen whether the ammeter is installed on the positive or negative side of a basic circuit.)

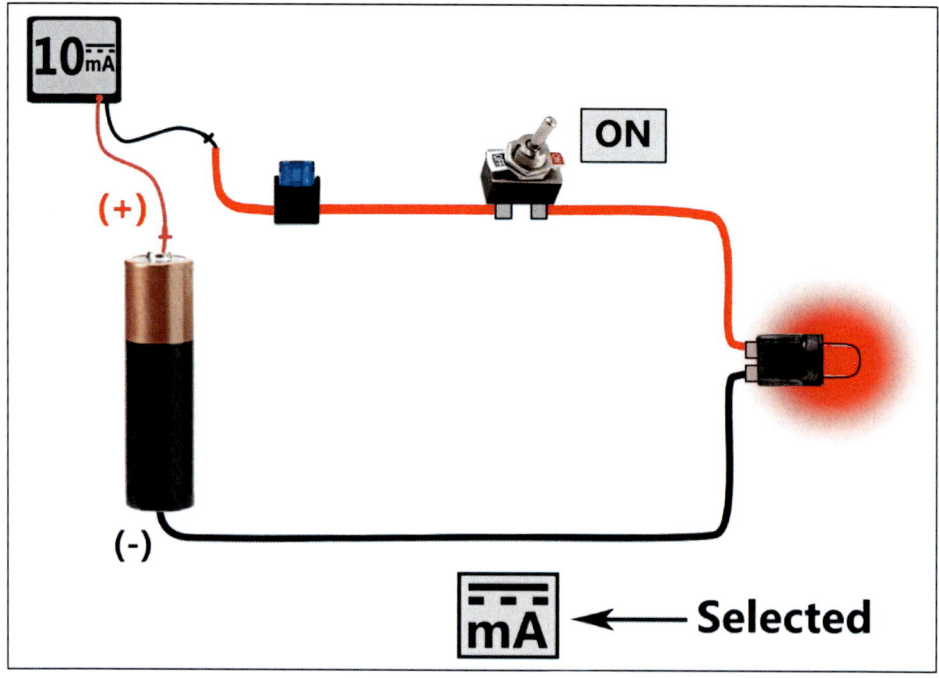

(This example shows how an ammeter is installed into a circuit to now read **milliamps**. In this example we decided to install the ammeter in the positive side of the circuit. Regardless of where we install the ammeter, we will always read the same current throughout this circuit.)

Milliamps is not a new unit of measure. It is just a smaller amount version of amps. We will only use this setting on very low powered circuits where we know that the electricity flowing is not going to be very much.

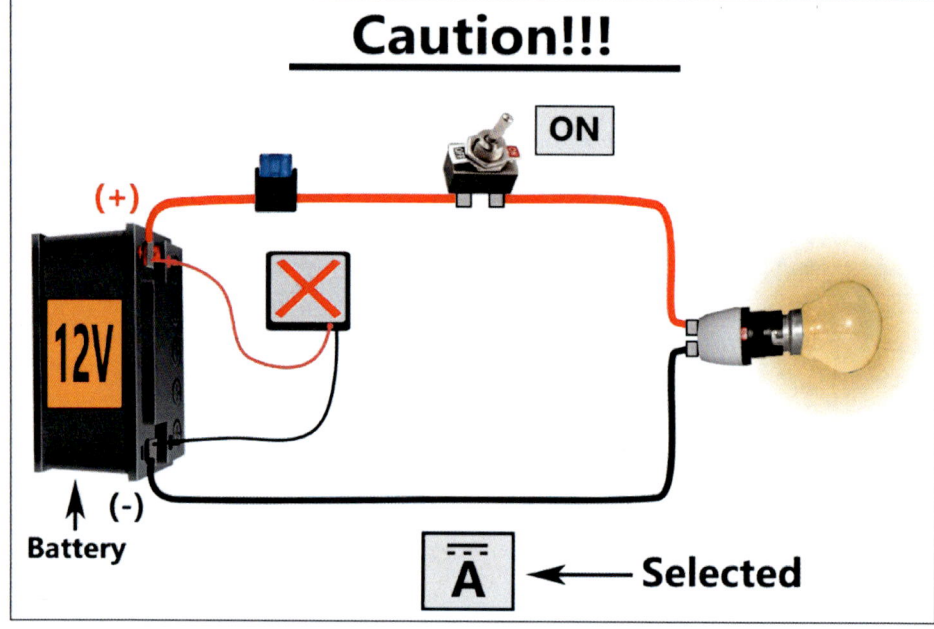

Remember that you must NEVER use an Ammeter across a battery or other power source as you will surely destroy or cook your meter. Think of the ammeter as a set of jumper wires with a gauge on it. You should NEVER put jumper wires across a battery. You would create an electric short from power directly over to the negative side of the battery.

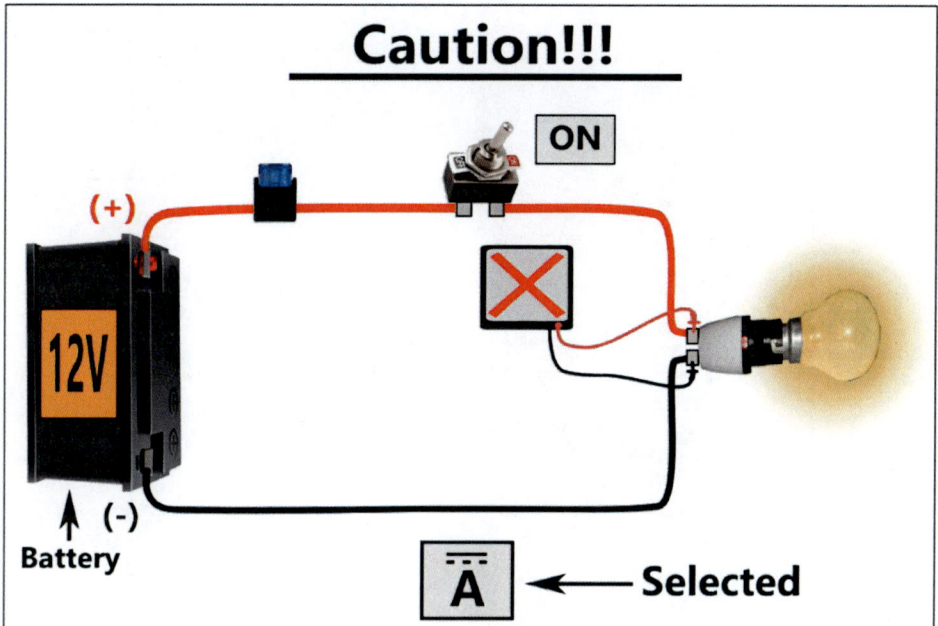

Another common beginner mistake, using the ammeter like if it were a voltmeter. Creating a short across the electrical load of the circuit. Remember that an ammeter is like a set of jumper wires. If you "jump" the battery or the electrical load you will create an electrical short. Never bypass a power source or an electrical load. Be careful not to make this mistake.

There is actually another tool available that reads amperage that measures in a way that makes it MUCH easier to use, safer and quicker in taking measurements. This requires the tool known as a clamp-on ammeter or "amp clamp". This tool makes taking amperage measurements a cinch without the fear of messing anything up by connecting it wrong like the regular ammeter.

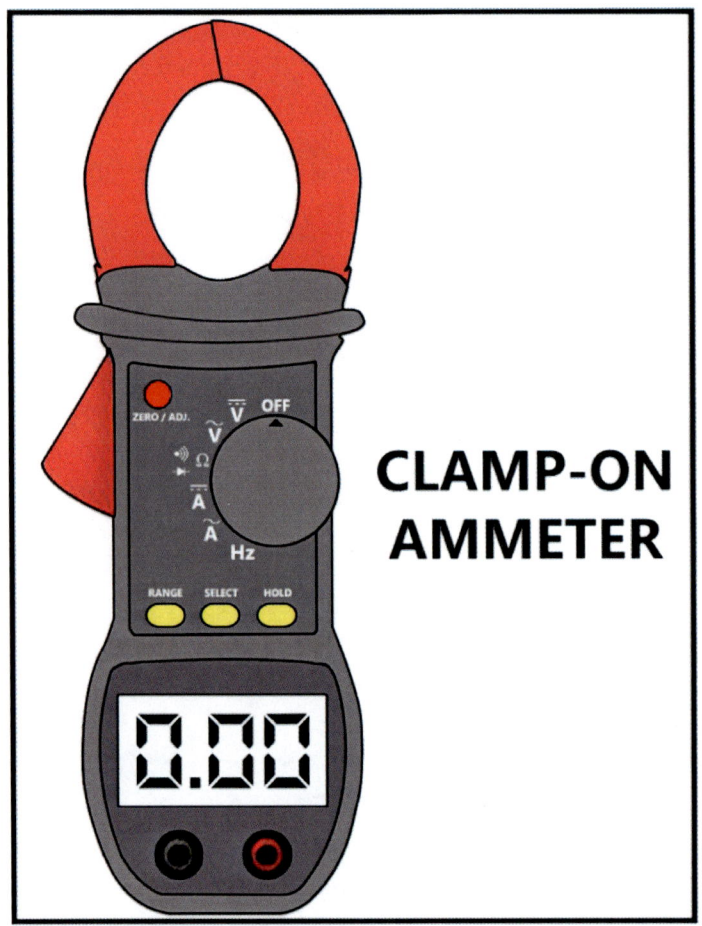

(This is an image of what a clamp-on ammeter looks like. This type of ammeter makes current flow testing as easy as opening the clamp by using the lever on top of the meter and clamping it onto a wire on the circuit. Remember that the circuit can only be tested for amps while the circuit is ON.)

This method of amp testing is superior because the measurement is easy to do, no damage can be done to the circuit or your meter and it takes a lot less time to perform a reading.

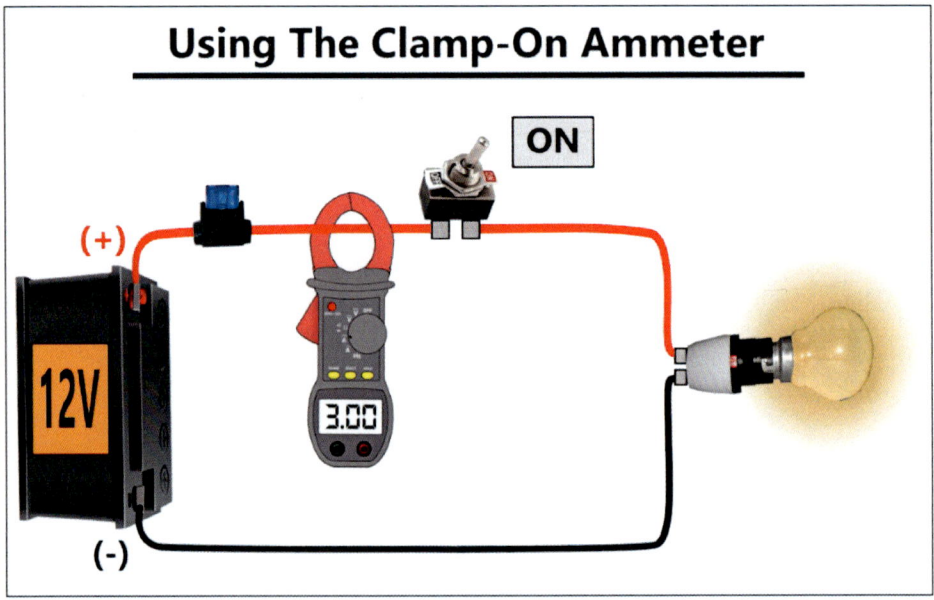

(This circuit is being checked for current flow using the clamp on ammeter. You can see how much easier it is to use than the other kind of ammeter. No disconnecting is required. Just simply clamp it onto either the positive or negative wire and turn the circuit ON.)

In this example, everything in the circuit is working normally and the clamp-on meter reads 3 amps. Please note that this reading is only there for educational purposes, do not take it as a baseline reading for a light bulb.

Resistance:

The last thing we need to understand for this book is electrical **Resistance**. This is anything that resists, restricts or slows down electricity from flowing properly throughout the circuit. It can be compared to a clogged part of a water pipe in a water circuit. The clogged piece of a pipe resists the proper flow of water. It can actually be compared to anything that would stop or slow the flow. **Electrical resistance** resists the flow of electricity.

Examples of Common Unwanted Electrical Resistances:

- An electrical wire with damaged, burnt, or missing strands
- Corroded or damaged electrical connectors
- Corroded or damaged electrical contacts and terminals
- Any loose or bad electrical connections

The tool used for measuring resistance is known as the **Ohmmeter**, or ohm setting on a multi-meter. Here are example of the ohmmeter measuring **Resistance**…

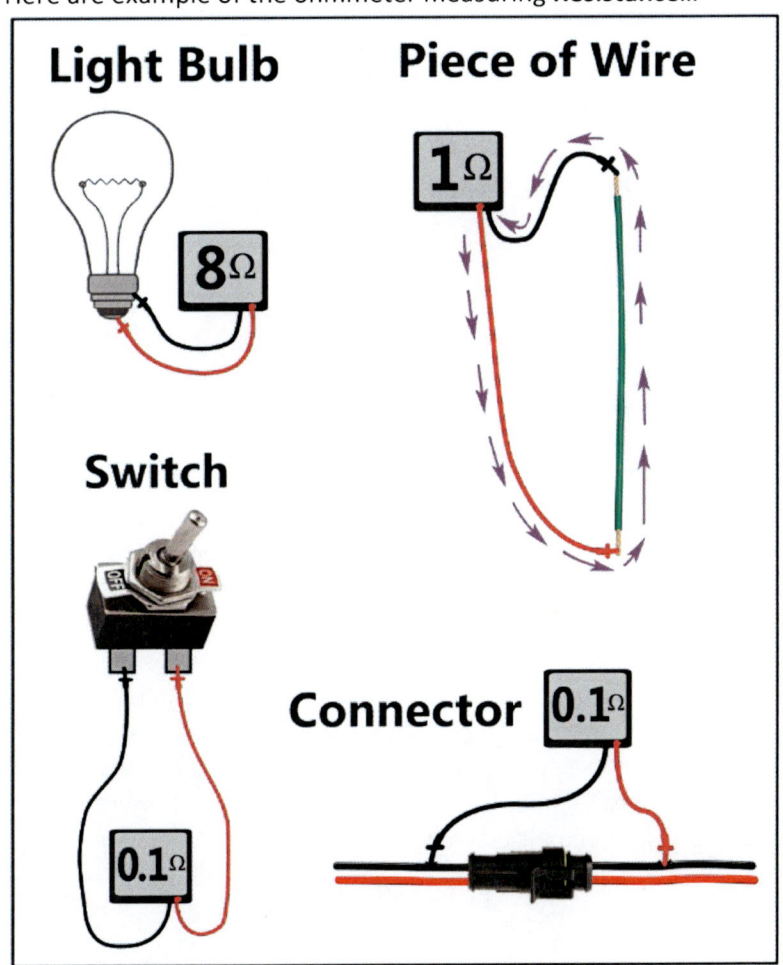

14

(Here we see various examples of the ohmmeter measuring resistance. You can see example ohm readings for a wire, an electrical device (this case a light bulb), a connector and a switch.)

These readings are not to be taken as something to follow as a baseline, they are only for education purposes. You may have noticed that when testing for resistance you are testing the part removed off the circuit. The only way to test a part for resistance is with the circuit off or with the part disconnected.

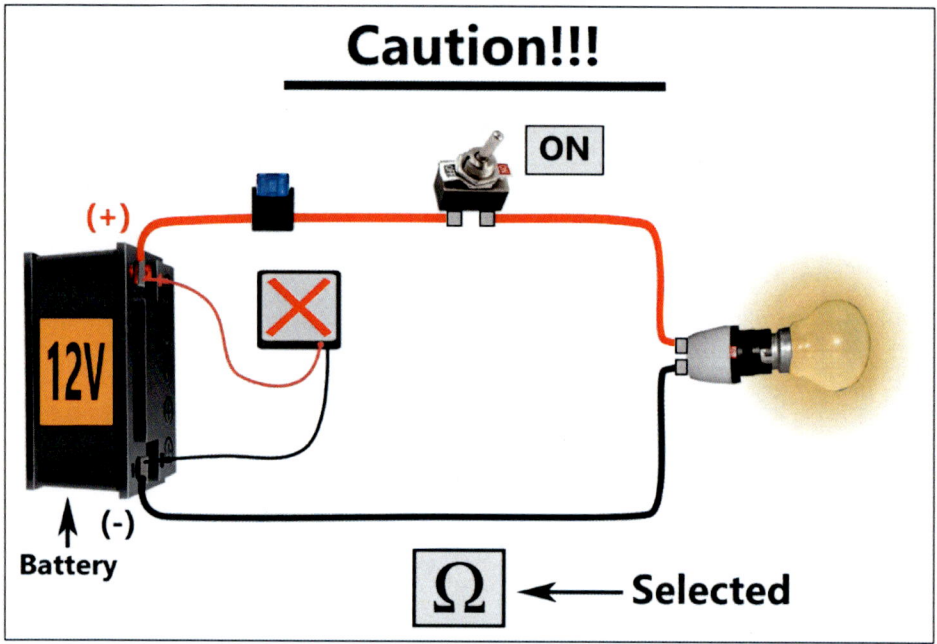

(Remember to NEVER use an ohmmeter across a battery as you will surely destroy your meter this way as well.)

Personally, I never use an ohmmeter for electrical troubleshooting. I consider it the worst tool for diagnosis. Although many may recommend it, I would never use it for diagnosing a problem. The test is better suited for bench testing more than actually troubleshooting anything. The best meter to have for tackling nearly all electric problems EVEN resistance problems is the Voltmeter and I will explain why later.

Important Note about Hidden Resistances:

Resistance of a component or wire can be measured using the ohmmeter but this is not a 100% accurate test. Did you know that even if a wire had only one good strand left in it and all the rest of the strands were gone or damaged, the ohmmeter would still read as if it were good in resistance? How can this be when clearly broken wire strands qualify as a resistance to electricity? Well the truth is, the ohmmeter only measures the resistance of the complete strands in the wire. If all the wire strands in a wire were broken, then the ohmmeter WOULD read an accurate reading of "OL" indicating a fully open wire or broken circuit. The problem with the ohmmeter is that it will not pick up any hidden resistances such as missing wire strands or loose connections. Be mindful of this when using an ohmmeter. The best way to truly find all resistances in a circuit is actually with a **voltmeter**. I will explain how to test for this in this book.

Another Note on Resistance of Electrical Devices:

Any electrical device itself is a Resistance to electricity. Whether it's a light bulb, a fan, a heater, a fuel injector it will have a very specific resistance in it. This resistance is the only resistance that should be present and acceptable in the circuit. Each electrical device has a resistance in "ohms" that has been taken into account by the engineers that built the circuit. The only time the resistance of a device is unacceptable is when it does not contain the correct resistance it should have as compared to when new.

Example: A light bulb has 8 ohms of resistance when new. When you check a used light bulb with the ohmmeter it has a reading of 15 ohms. If it is the exact same part number as the new one, then this light bulb has too much resistance in it as compared to when new. It will likely not shine as bright either.

Important Reminder On Replacing Any Electrical Part:

When replacing any electrical device or modifying a circuit you should check to see that the replacement electrical part is around the same resistance reading as the original. DO NOT install any replacement part that is a LOWER resistance reading than the original part as this can likely blow a fuse or even damage the circuit. Why, you ask? Well the less resistance, the more electricity flows in the circuit and the closer the amount of current flowing gets to reaching the maximum electricity, in amps, that the fuse is rated to "blow" at. Always replace a device with one that is very near or slightly higher in ohms than the original but never with one that is significantly lower in ohms.

Final Note about Resistance & Temperature Changes:

When testing for resistance in a circuit using the ohmmeter, you have to turn the circuit off and then ohm test the piece with it disconnected from the circuit. I will tell you why this is a bad idea as well as very time consuming as far as troubleshooting. First, you are required to waste time disconnecting things to get a resistance reading that is likely to be inaccurate many times. Then second, you might actually damage or break a part of the circuit when doing this if it is forced apart accidentally. Did I mention you will take forever to find a problem using the ohm test?

The BIGGEST problem with resistance testing and using the ohmmeter for professional diagnosis is that the circuit can begin to fail only when it is either hot or cold. It may short out or maybe build up too much resistance when it is ON but then when it is turned OFF, it goes back to having a normal resistance once it has cooled down. Keep your ohmmeter only as bench tester for replacement parts or spare components as it will not work well for diagnosis.

Now do you see the flaws of the ohmmeter? Don't get me wrong it works great as a bench tester for loose or replacement parts, but not for troubleshooting a problem. Instead throw your ohmmeter back in your toolbox and get out your trusty voltmeter. The only way to TRULY test a circuit is when it is ON and working.

Now that all those refresher lessons are out of the way let us finally take a look at some examples of testing..

How To Test Circuit Protection Devices:

The method of testing both a fuse and a circuit breaker is the basically same way. The test involves connecting a voltmeter across the terminals or the wires on each side of the circuit protection device while the circuit is ON. Depending on the style of device being used in the circuit, it may be necessary to piercing the wires if unable to access the terminals. Here is how you test them...

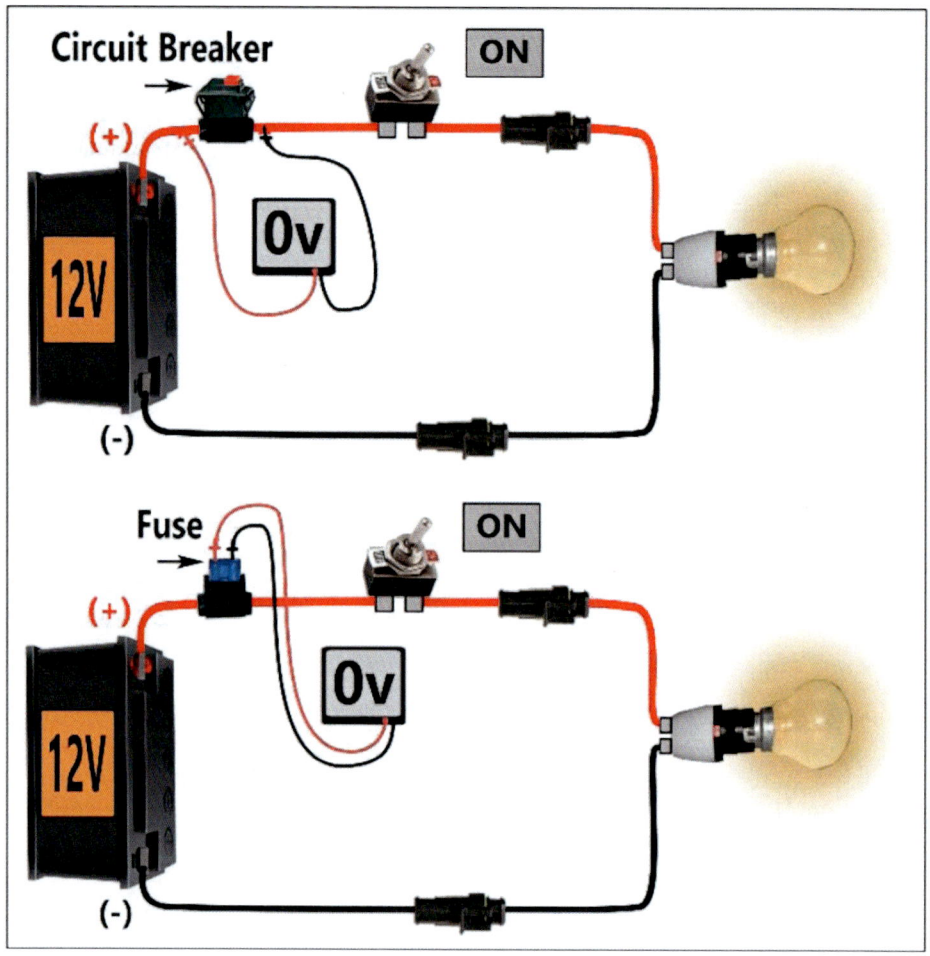

(In this example we see how a circuit breaker and a fuse can be tested in the same way. The voltmeter is installed to test the voltage difference across the two sides of the circuit protection device. A circuit breaker or fuse should have **near** 0v across it when tested by a voltmeter and the circuit is ON.)

Some fuses have little test slots on the top side or open terminals that can be used as test points as shown. A reading of near 0v across is considered a good reading for any circuit protection device. Any voltage reading that is high or the same voltage as the battery indicates that the fuse is blown or that the circuit breaker has been tripped.

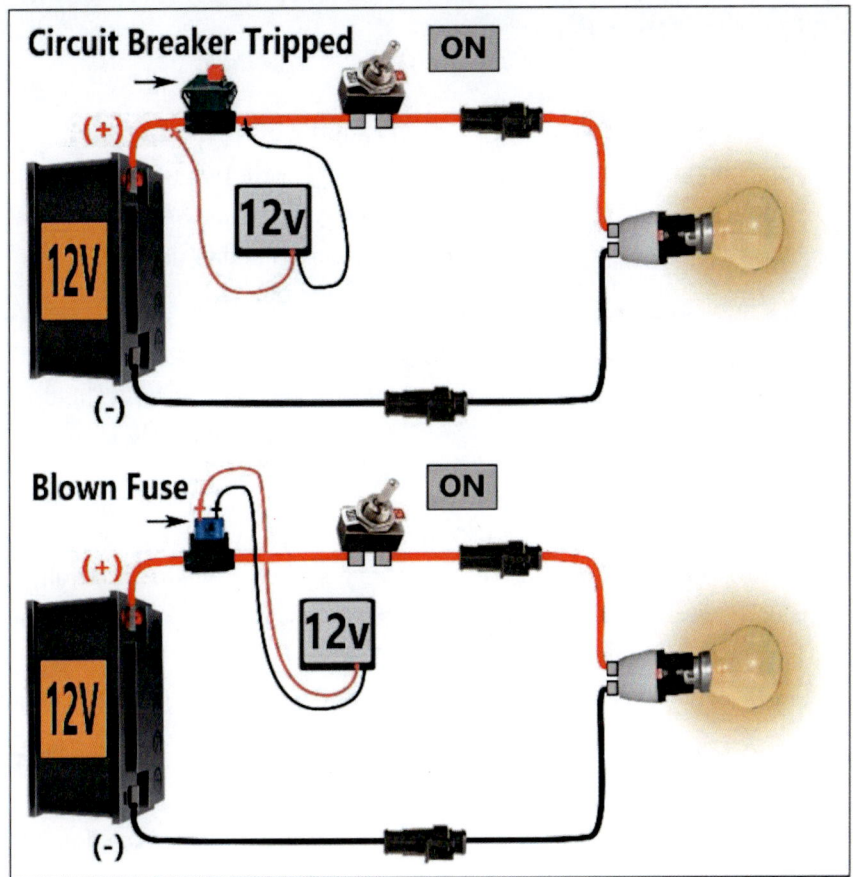

(This illustration shows the previous circuits, but now with the reading of a tripped circuit breaker and the reading of a blown fuse. The readings across them is 12v, in this case the power source's full voltage.)

A voltage reading that is equal to the power source voltage does not always mean that the problem is the fuse or circuit breaker, as it rarely is. The problem can be somewhere else in the circuit. It might be an electrical short in the circuit that caused the fuse or circuit breaker to open. In the example for the circuit breaker, we can reset the breaker to see if a short exists. If the breaker trips again after resetting, then an electrical short is for sure the problem. In the example for the fuse, we will have to replace the fuse to see if a short exists. If the fuse blows again then this is also a sign that there is a shorted circuit problem

In either case the problem may not always be the circuit protection devices but instead something else that caused them to fail. A circuit protection device can fail from an overload in current or from a power surge in the circuit. The general takeaway from this test is to understand that any fuse or circuit breaker should have near 0 volts across it with the circuit ON. If a protection device has a significant amount of volts across it this is proof that a problem exists. If you are interested in more on this subject, please check out "Everything Electrical: How to Find Electrical Shorts" for the complete and thorough explanation on how to solve a shorting problem.

In the next chapter we will finally begin to see more of the testing portion of the book. After enduring the review of the basics of electricity we are now ready to begin…

Ch2: Voltage Testing Like a Pro:

Introduction:

Voltage inside of a circuit is supplied when a power source that produces voltage or a power source that stores voltage is connected to the circuit. There are two main kind of power sources you will come to see in the electrical world. One type of power source creates electricity while the other type of power source can only store electricity.

One example of a power source is a **generator**. The "generator" produces voltage (electrical pressure) when it is ON and spinning. When the generator is connected to a circuit, it is able to supply power to the electrical device of that circuit. This is the way households and many industrial places power their electrical devices. There is a large generator at the electrical company that supplies all the electricity to power multiple homes.

Another example of a power source that provides voltage (electrical pressure) to a circuit, is a **battery**. I'm going to assume you know what a battery is since they are practically everywhere including inside of cameras, phones, laptops and all other portable electrical devices. A battery actually does not create voltage like the generator, it just stores it and when connected to a circuit it can power the electrical device in the circuit for a limited time.

There are two kinds of voltage. **AC voltage** (produced by an AC generator) and **DC voltage** (the voltage stored inside of a battery). Although they are very different, regardless of the kind of voltage that the circuit is working on, voltage testing is pretty much done the SAME WAY. The only difference you will notice when testing for voltage in a circuit is the dial position you will choose on your meter. The voltage setting selected should match the kind of power source in the circuit.

Shock Hazards Safety:

Whether its AC voltage or DC voltage that you are working with, the testing method is essentially the same. The problem is when the voltage level in a circuit is above 48 volts there is a possible danger. Anything that is above 48 volts is very hazardous and can potentially kill you from an electric shock. Household voltage is 120 volts AC so be careful when working with this voltage and put on your insulated electrician's gloves to protect you against a shock.

Examples of high voltage circuits where you should ALWAYS use your insulated electrician's gloves include…

- High Power Industrial Circuits: Electrical company circuits, large electric motor circuits, Power lines, etc. Which can range from 480 volts to many tens of thousands of volts!!!
- Hybrid and Electric Vehicles: Which range in voltages from 200 volts to up to 700 volts or more.
- Metro or Electric Train Power circuits and power lines. Which can easily go up to 36,000 volts

This sort of voltage is nothing to take lightly because it can fry you in an instant. Please be very careful and heed my safety warning about using your insulating safety gear when working with higher powered circuits.

Now then, let us begin using your voltmeter like a pro..

How It Really Works

As I mentioned before, voltage testing is the superior way to test many electrical problems, so it only makes sense that I focus on covering how voltage "moves" in a circuit. Remember that voltage is like the water pressure in a water pipe circuit. Voltage is electrical pressure and it is what makes electricity flow. In this chapter, we will see how voltage itself moves during various circuit conditions. But in order to understand how voltage moves we need to first review how the voltmeter REALLY works so that when I begin explaining other lessons you are not lost or misunderstand anything that I say.

Review about the Voltmeter: Remember that voltmeters DO NOT read voltage. They actually read the difference in voltage levels between the two test probes. It takes a sample of the voltage that is present at each of the test probes and then subtracts the values and displays the subtracted answer on the screen.

Example 1: If you had one test probe on the negative battery post (which has 0v) and then the other probe on a terminal that had (24v), then the voltmeter would display the difference. In this case 24v-0v is 24v.

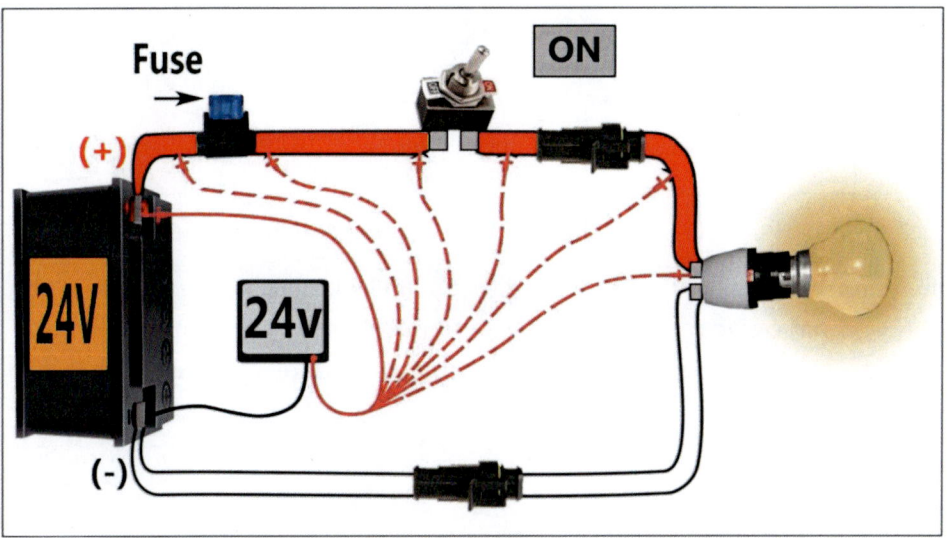

(Picture of a circuit being probed in multiple areas on the power side of the circuit. One of the voltmeter probes stay at the negative terminal or ground, while the other probe is used to test various points for voltage on the power wire. All of the areas tested are places where voltage SHOULD be present in a normally operating circuit.)

Notice that whenever you have one probe attached to negative, you can use the other probe to check for voltage at various places in the circuit. Because one probe is placed on the negative (0 volts) the voltmeter reading will display the actual voltage available at the location of the second probe.

Important Note About The Illustrations: You may have already noticed that the circuit illustrated in the previous image shows the top wire being colored in red. This does not mean that the wire is red in color. The red actually represents where voltage is present in the circuit. By making the wiring see through and coloring in parts of the wiring red we can better understand how voltage moves throughout the circuit during various circuit conditions. Later you will understand why I have done this.

Example 2. Voltage Differences: If you had one probe placed on a terminal that had (9v) and the other probe on a terminal that had (3v), what would the voltmeter display? The voltmeter would display the difference in voltage between the two test points, so it would display 6 volts. (9v-3v=6v).

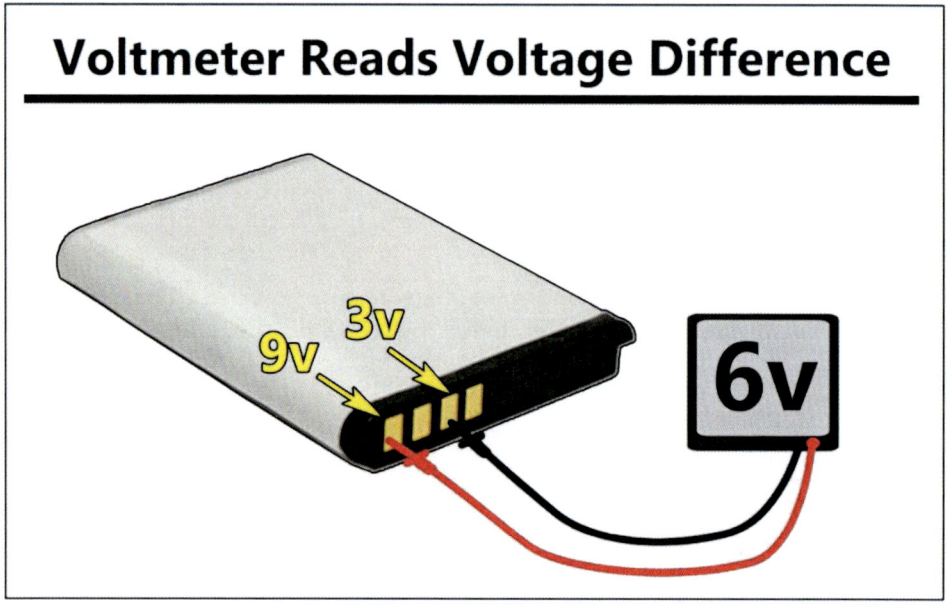

(Example of what the voltmeter will display when probing two points with different voltages.)

Example 3. Similar Voltages: If you connected both the voltmeter probes to a part of a circuit that had (24v) what would the meter read? Well if BOTH the voltmeter probes were connected to somewhere that had the same 24v then the voltmeter would read 0v (24v-24v=0v).

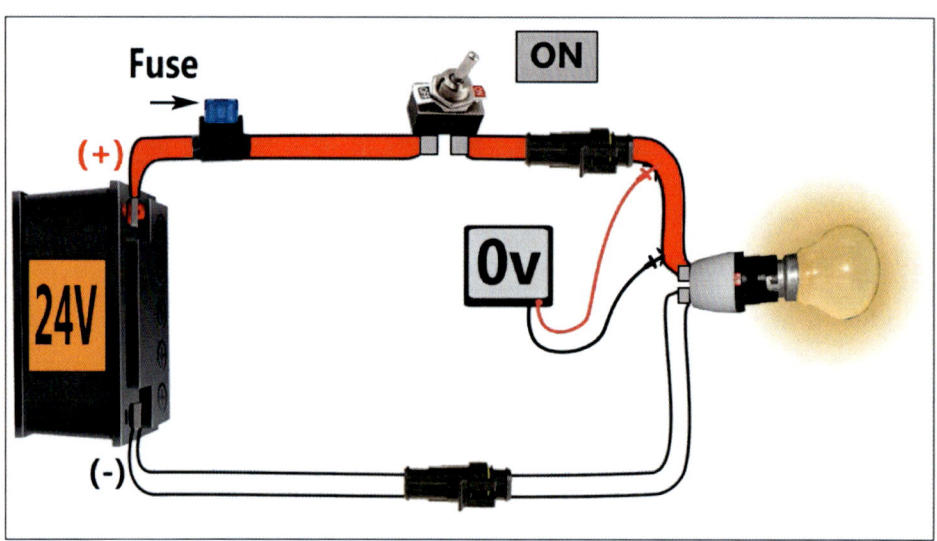

(Illustration of a voltmeter probing two points on the power wire. These two points have the same voltage inside them so the voltmeter reads 0 volts.)

These examples are meant to teach you how the voltmeter arrives at the reading it displays. This is important because there are actually two ways to use the voltmeter. One way is to test for voltage by connecting one probe to negative (0 volts) and using the other test probe to probe a place in the circuit you would expect to have voltage at. The second way to use the voltmeter is to NOT place one probe to negative, but instead place the test probes across an electrical part. This would read the difference in voltage between the two test points.

Now let us see the circuit in various conditions and note how voltage moves….

How Voltage Moves:

Voltage does NOT flow like how amperage flows, but voltage does "move". By learning this concept, you can assume or predict where voltage is going to be in the circuit. Let's take a look at how voltage moves in a circuit during different conditions.

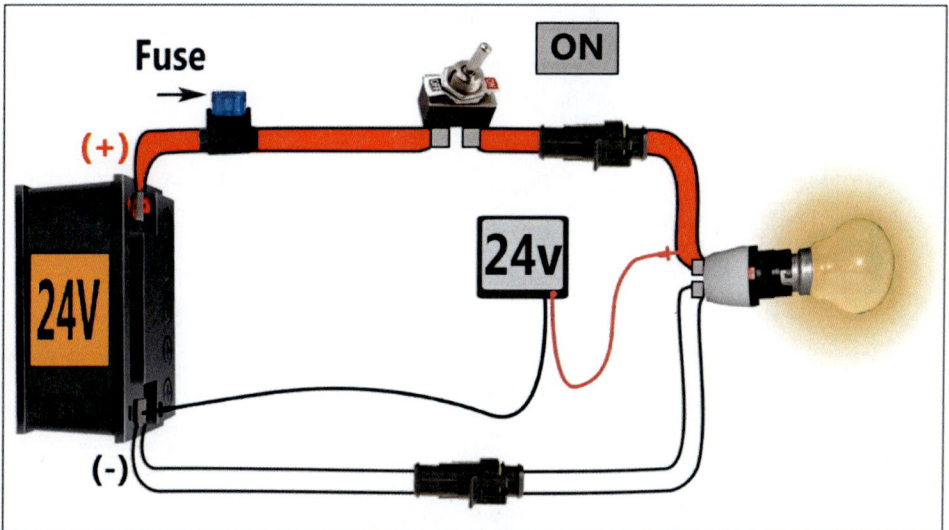

(This is our circuit operating normally. You may have wondered at some point why voltage stops at the electrical device. Well the truth is, the electrical device uses up this voltage to convert it into something useful such as light or movement or heat. That is why there is little to no voltage left on the negative wire of a normally operating circuit. All the voltage is used up by the main resistance in the circuit, this case the light bulb).

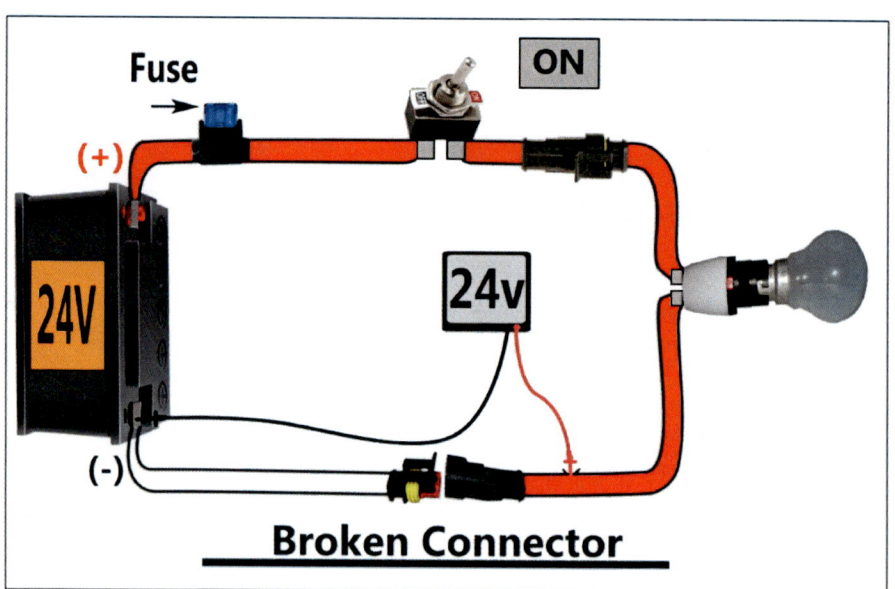

(This illustration shows the circuit not working because of a broken or disconnected connector on the negative side. As you can see the voltage moved through the light bulb and to the point where the circuit is broken. The voltage is able to freely move to the negative side wire because the light bulb is not able to convert the voltage into light anymore because the circuit is not complete and no electricity is flowing. The voltage is able to just go in and out of the electrical device without being affected.)

Note: Whenever you have a broken circuit, the voltage will always move to be across the open or broken part of the circuit. The trick is to probe the circuit in various places to find where the missing voltage is.

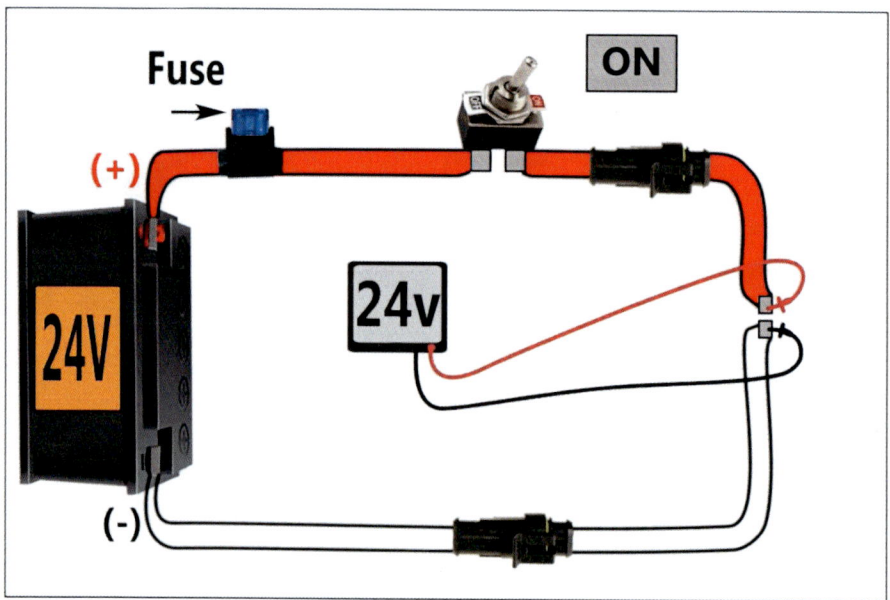

(This image shows the light bulb disconnected and how the voltage is present up to where the circuit is open or broken. Checking for voltage at the electrical device is also a common first step to diagnosis of any problem. It will tell whether you have voltage up to that point or if you are missing voltage.)

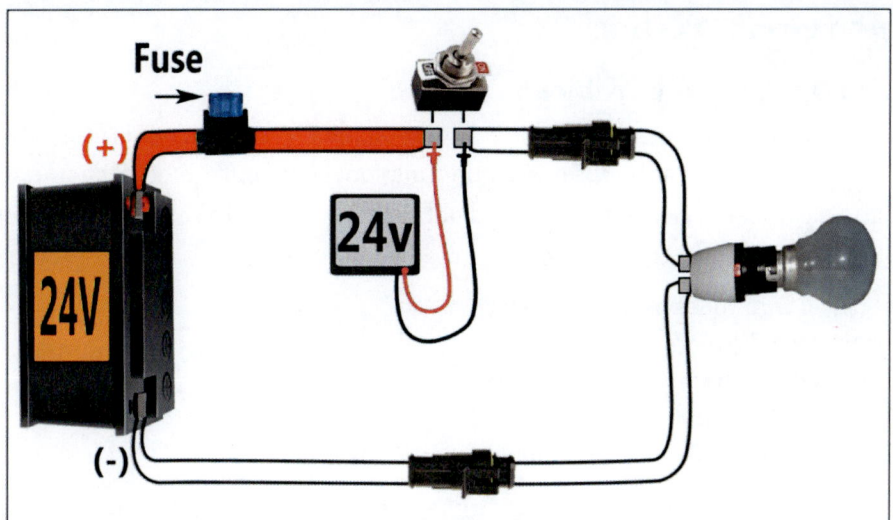

(This illustration shows how voltage moves when there is a broken or disconnected switch. The voltage is present across the open part of the circuit, the switch. If the switch where disconnected and a voltmeter was placed across the wires of the switch, it would read the battery voltage or in this case 24v.)

Note: This is actually how you can also test a switch to see if it is good or bad. You put your voltmeter across an installed switch and if it reads battery voltage in both the ON and OFF position, the switch is BAD and not letting the voltage move through it on its way towards the electrical device.

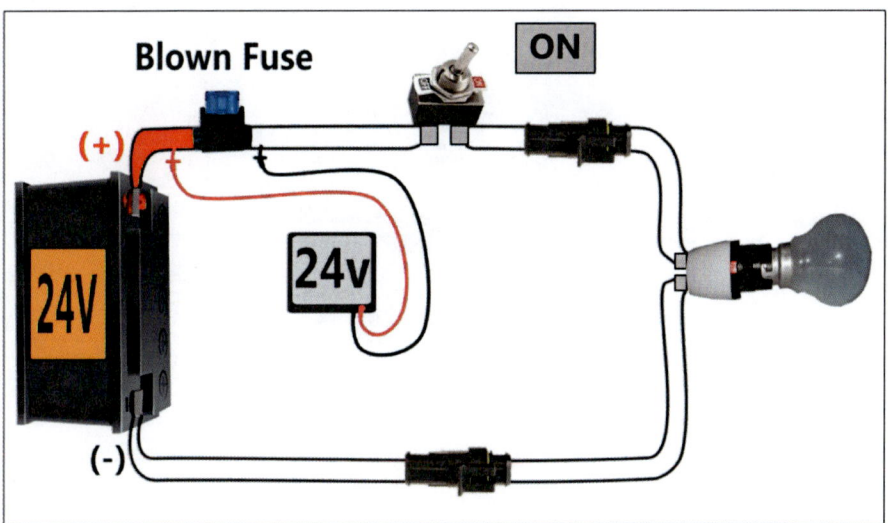

(This illustration shows where the voltage would be if a fuse was blown or disconnected. The voltage again is present at the open or break in the circuit closest to the positive battery post.)

By now you should have a good idea about how voltage moves and be able to predict where it might be present when you have a problem with an open circuit.

Ch3: Open Circuit Testing

Signs of an Open Circuit problem.

When a problem with a break or open in the circuit happens, you will usually see the symptoms of the problem in the form of a non-functional light bulb, motor, heater, etc. Basically, the electrical device in the circuit is not working. Here are some examples of common problems that are usually the cause of it.

- An electrical wire that has been broken
- Broken electrical connectors
- Broken electrical contacts and terminals
- Any bad electrical connections
- The electrical device itself being open internally

The steps to take for finding an open is fairly straight forward. The voltage in the circuit will be where the open or break is. All you have to do is find the last place where there IS voltage and that will lead you right to the open.

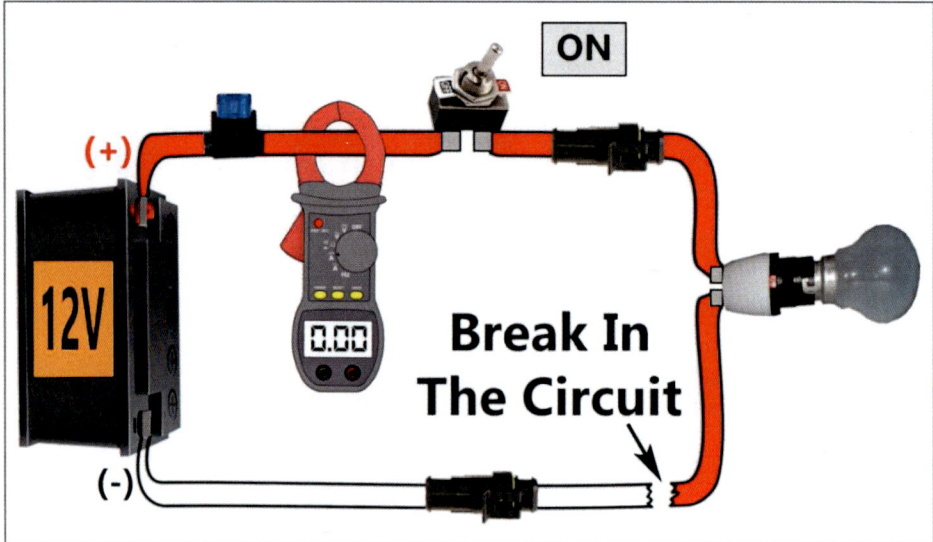

(In this image, we can see an amp clamp being used on a circuit that isn't working. By checking for amperage we can quickly determine if the problem is in fact an open circuit.)

This is what I always do first whenever I suspect an open circuit problem. I turn the switch for the circuit ON and clamp onto one of the wires with my clamp-on ammeter and check for amps. The 0 amps reading confirms that there is in fact an open circuit somewhere.

If you do not have a clamp-on ammeter yet, another way to determine if you have a problem with the circuit is to probe across the electrical device with the circuit ON. A 0v reading at the electrical device would confirm that there is a problem in the circuit.

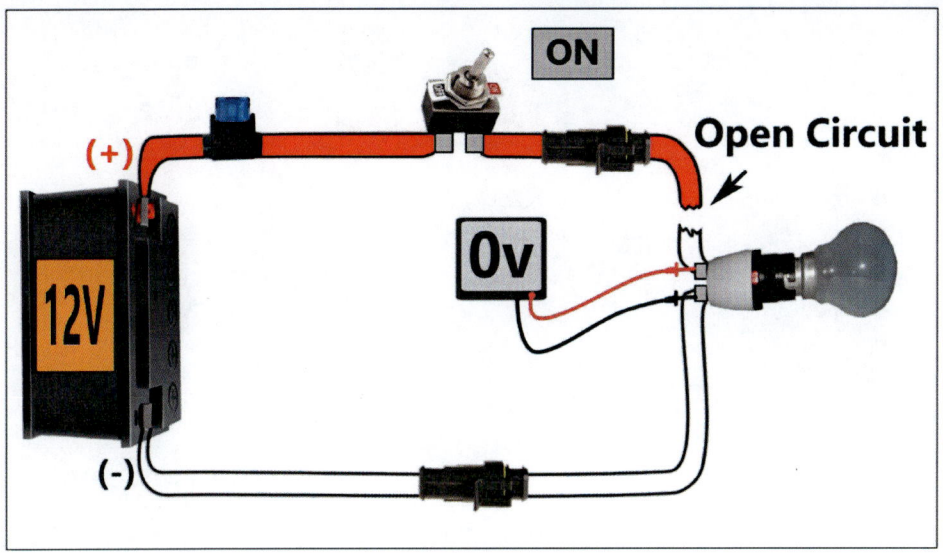

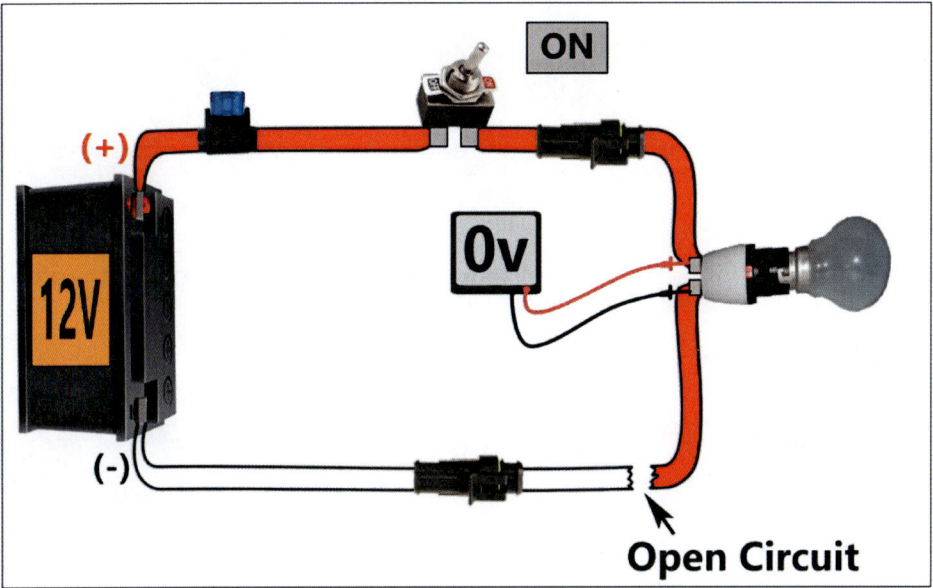

(Two examples of an open circuit being confirmed using the voltmeter at the electrical device. The problem can be either in the power wire or the negative wire. Regardless of where it is, the reading at the electrical device will be 0v.)

So how do I know which wire is the one with the open in it? Simple. Connect one test probe of your meter to negative, disconnect the electrical device and probe both sides of the electrical device wires with the second probe.

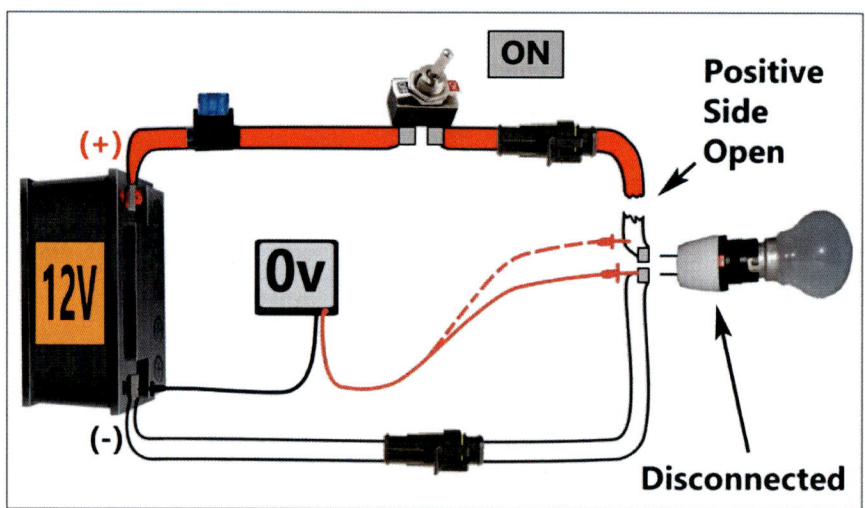

(Disconnect the electrical device, place one test probe to the negative side of the circuit and use the other test probe to test each side of the electrical devices wires. If you read 0v on both sides you have a Power Side open circuit. A reading of zero on both wires proves that there is no power coming in to the electrical device.)

Now lets that a look at an example of a negative side open circuit.

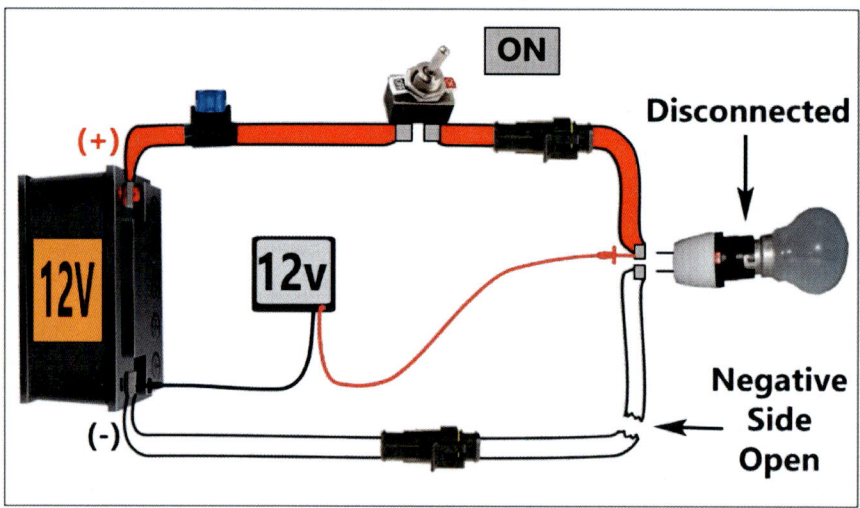

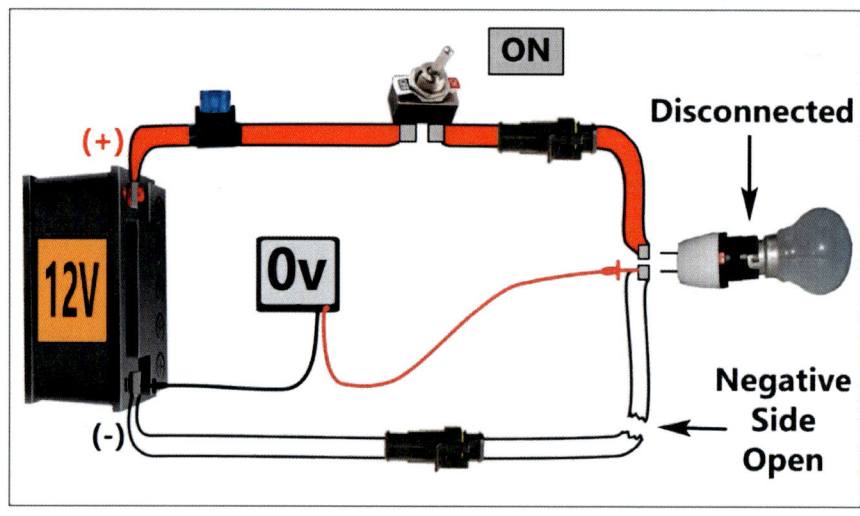

(Start by disconnecting the electrical device, place one probe to negative and use the other probe to test each side of the electrical devices wires. If you read full voltage on one side, this case 12v, but then 0v on the other side you have a Negative Side open circuit. The wire with the 12v is the power wire of the circuit.)

By performing this voltage test, the results will tell you whether you have an open circuit in power side or the negative side of the circuit. Now that you know what side the open is on, it's only a matter of narrowing in on the "open".

There are actually two different methods to take depending on where the open is located in the circuit as we will see in the next sections. Remember to always perform voltage testing with the switch turned ON in order to diagnosing the problem properly.

For the Power Side Opens:

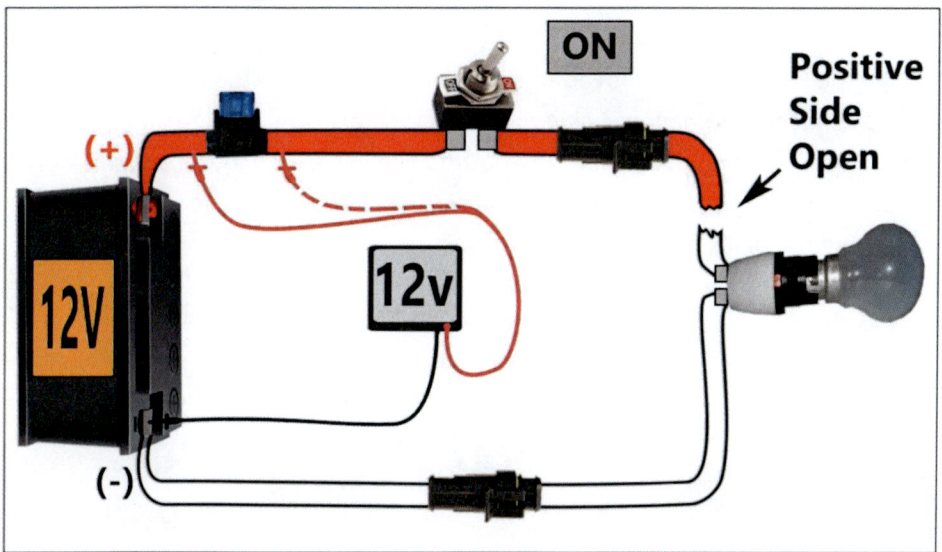

(The first step to take to finding a power side open is to connect one test probe to a part of the negative side of the circuit and then use the second test probe to check for voltage on various points in the power side. In this image we are checking both sides of the circuit's fuse. If there is voltage present on both sides, this lets you know that there is nothing wrong with the fuse and the wiring from the fuse going back towards the battery. Also notice that the light bulb for the circuit has been reconnected after finishing the previous testing to find the side that the open was on.)

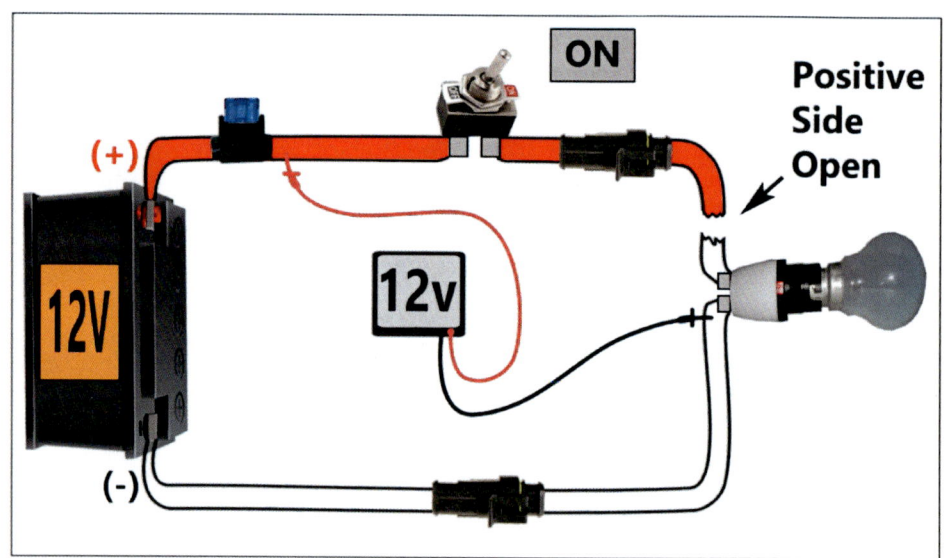

(In this image the black probe is moved from the negative post to the wire on the electrical device that is connected to negative side of the circuit. This is the same as if you were to connect directly to the negative post of the battery like in the previous examples as long as nothing is disconnected or bad on the negative side of the circuit. If the voltmeter still reads 12v after moving the black probe from the negative post to the negative wire on the light bulb, then we know that the parts between where the black probe was before and where it is now are all good.)

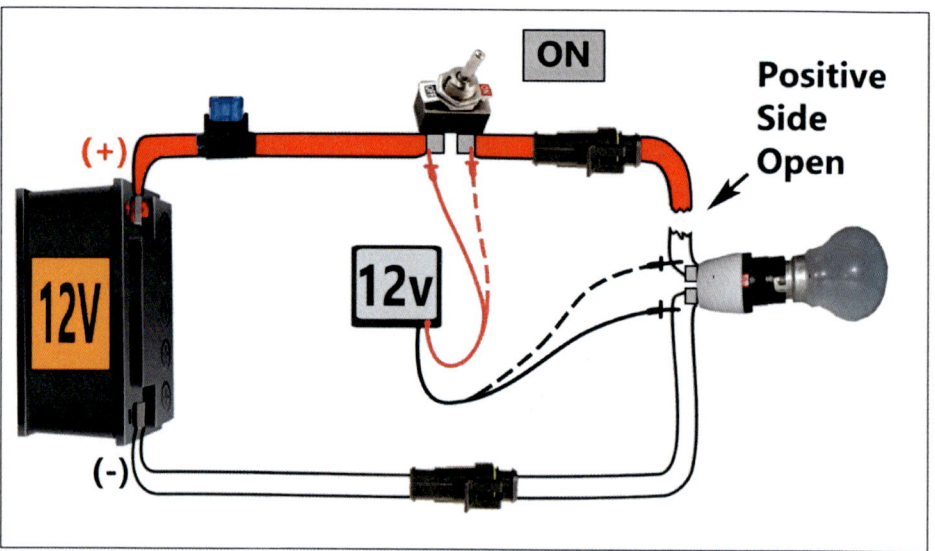

(The next step would be to check for voltage on both sides of the next part in the circuit, this case the switch. If there IS voltage on both sides of the switch while in the ON position, then you know the switch and all wiring back to the positive side of the battery is good up to that point. The only thing left that we haven't checked is the wiring between the switch and the current location of the black probe.)

Notice how in the previous image the black probe was also moved to the other side of the light bulb. This was done to narrow in on the open circuit. At this point we can narrow in by moving the black probe closer to the other probe on the circuit. If the voltmeter still reads 12v, then we know that the parts between where the black probe was before and where it is now are all good.

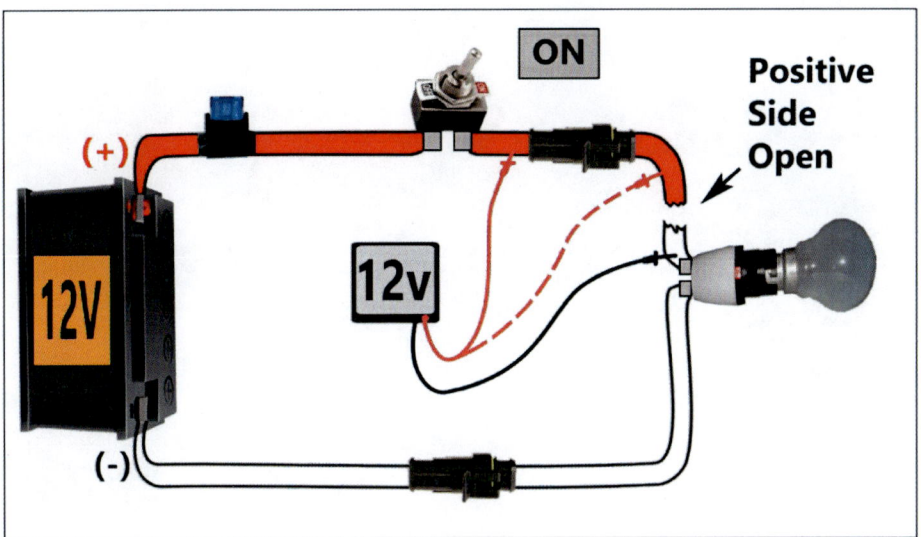

(The final step is to continue to probe the circuit, moving the probes closer and closer to each other. The voltage is still present before and after the connector so everything is good up to that location in the circuit.)

By this point you may be able to visually see where the break is. Once you do find the open it is simply a matter of repairing or replacing the broken part or area.

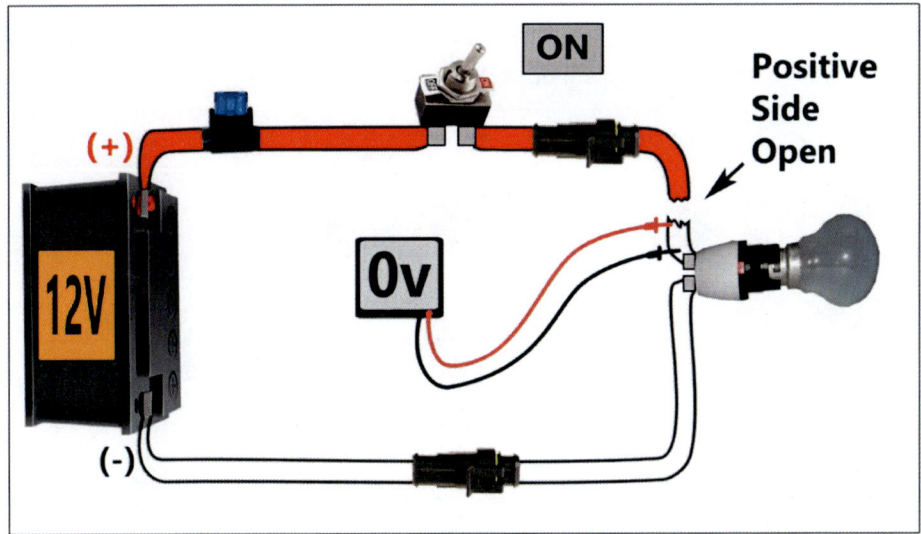

(If while narrow in on the open circuit, by moving the probes closer to each other on the circuit, you find that the meter suddenly reads 0v, this indicates that you have passed the area where the open is.)

Move the probes back to last place you still read battery voltage and inspect the area until the open is found. I included this extra illustration because some opens may not be easily seen right away no matter how near you are to it. They may be staring right at you the whole time and you may not visually see it. Luckily, the voltmeter has your back.

For the Negative Side Opens:

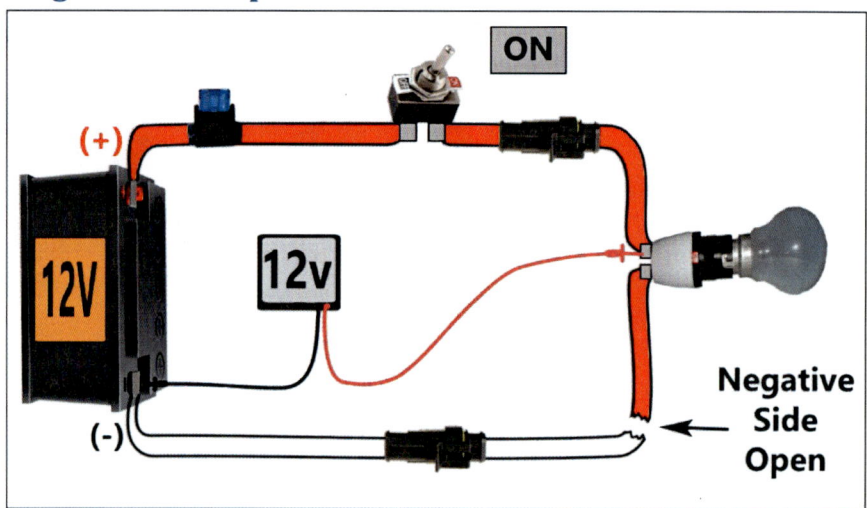

(In this image we continue from our initial test where we found that the open was on the negative side of the circuit. First you should note that the black probe is connected directly to negative side of the battery. The second probe is connected to the wire of the electrical device where we read 12v during our first test. Also, the light bulb has been reconnected for the remainder of our testing. This is our starting point.)

The next step to finding an open is to use the second probe to check for voltage in the negative side between the two test probes. If there is voltage present, this lets you know that there is nothing wrong with the wiring up to that point.

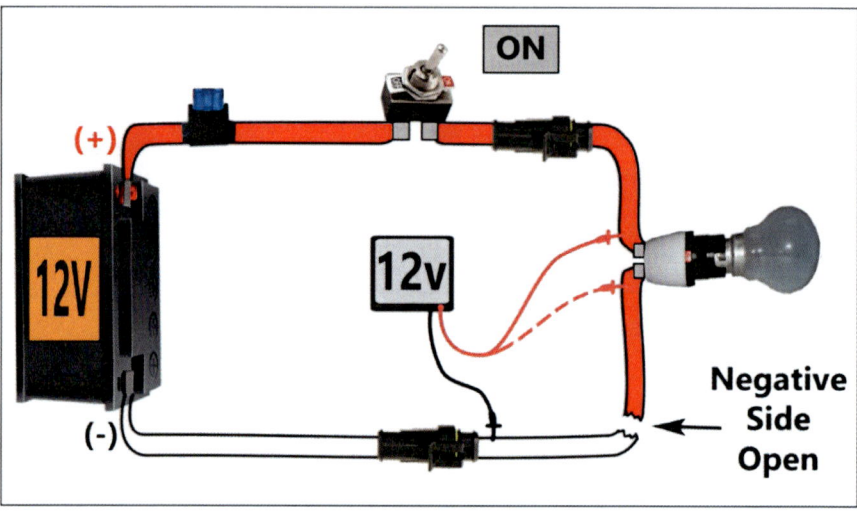

(We continue our testing by checking for voltage on the other side of the light bulb. If the other side of the light bulb also has the same reading, this tells us that the area from where the red probe was to where it is now is OK. This includes the terminals and contacts of the light bulb being good. As a second step, you can also move the black probe closer to the other red probe. If the reading is still battery voltage that means all the wiring from where the black probe was before to where the black probe is now is all OK. The only thing left is to check the wiring between where the two probes are probing the circuit.)

At this point you can begin to narrow in on the open by moving the probes closer to each other on the circuit. If the voltmeter still reads 12v, then you know that the parts between where the probes were before and where it is now are good.

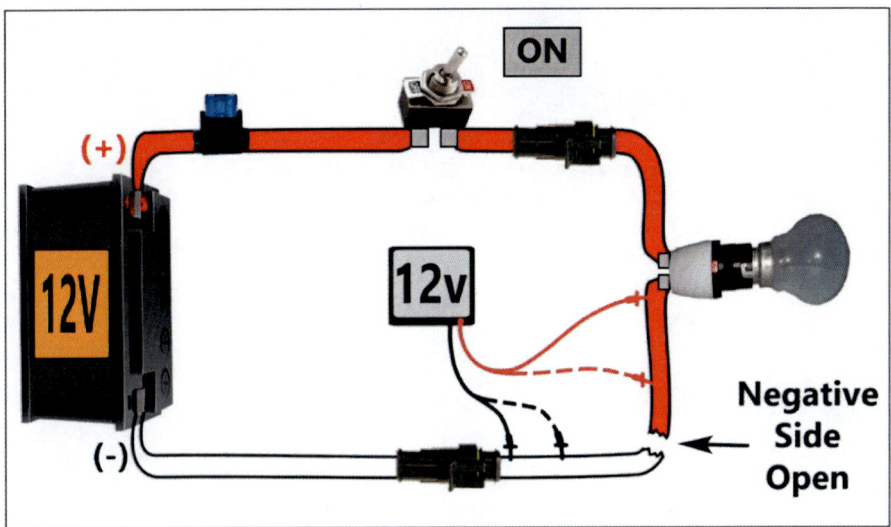

(The final step is to continue to probe the circuit, moving the probes closer and closer to each other. You may likely at this point be able to see where the break is. Once you find the open it is simply a matter of repairing or replacing the broken part or area.)

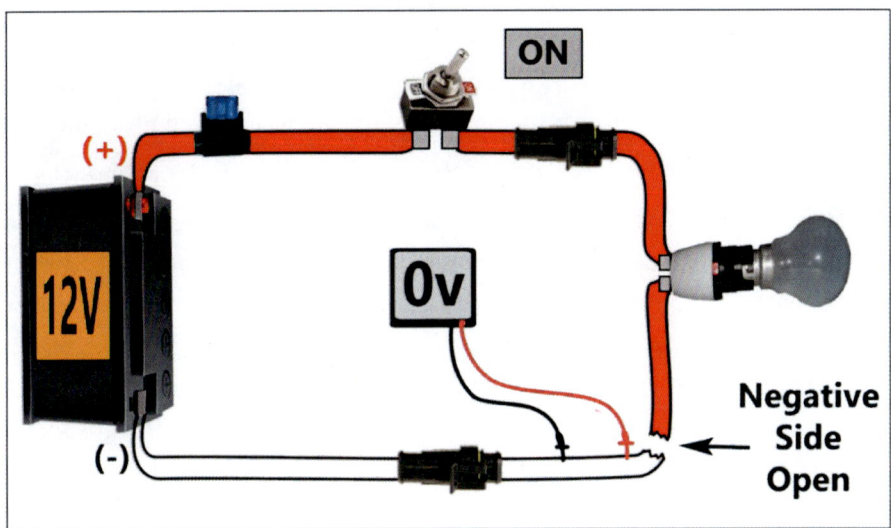

(If while narrow in on the open circuit by moving the probes closer to each other, you find that you suddenly read 0v, this indicates that you have passed the area where the open is. Move the probes back to last place you still read battery voltage and inspect the area until the open is found.)

Keep in mind that an open circuit can be anywhere. It can be a bad switch, a wire, a bad connector, broken contacts or terminals, the fuse, the battery or power source clamps and even the electrical device itself. Regardless of the location, what you have learned about voltage and how to use your voltmeter should be enough for you to understand what is happening with whatever open circuit problem you face and the steps you should take to solving it.

A Note on Multiple Open Circuits: Notice how I've mentioned that "the voltage is present at the break or open part of the circuit." If you notice in the following picture, I left the light bulb and the switch disconnected, in addition to the fuse being removed. These are both considered opens. If you were to put a voltmeter on them, you would read 0 volts across these "opens" or breaks in the circuit.

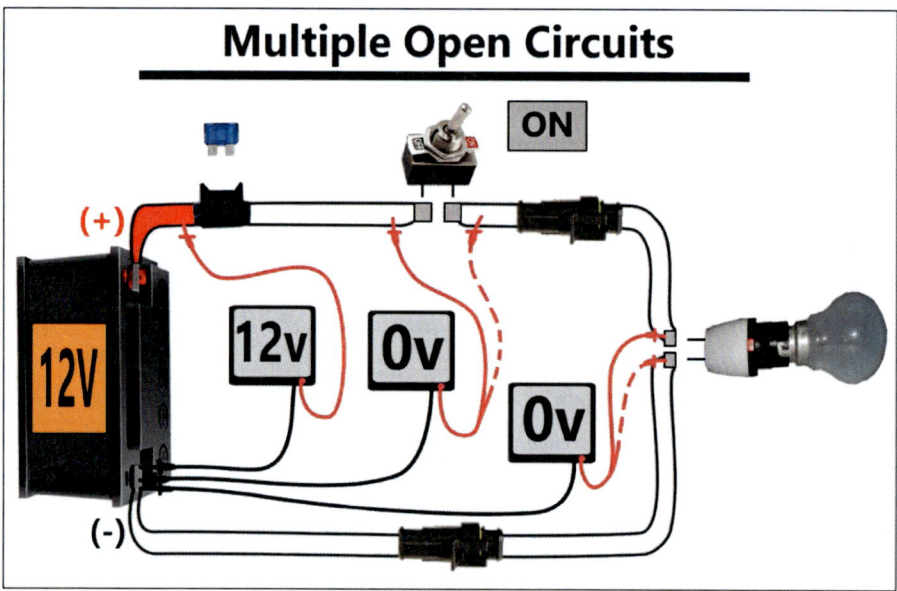

(This situation involves multiple open parts or breaks in the circuit. If you were to test this circuit in real life and needed to solve the problem, this situation would probably confuse you as I have already said that the voltage will always be present across the open.)

Well let me rephrase what I said by saying that **the voltage will be present across the open "CLOSEST" to the positive side post of the power source**. When you are dealing with more than one break in the circuit you will need to remember this. You will know there is more than one open if you read 0v across an obvious open you have already found. The opens you find should be tested for voltage, repaired and then if needed you should continue voltage testing to find the other opens in the circuit. The open that is closest to the positive post in the circuit with have all the voltage across it. That is how you will know it is the "open" closest to the positive battery post.

Tip About Fixing Open Circuits Faster: An open circuit can be quite time consuming sometimes when you have a very long circuit, you don't know where the circuit leads or you are just unsure of the area of the problem. Here is a helpful tip that can make the troubleshooting or repair a lot easier on you.

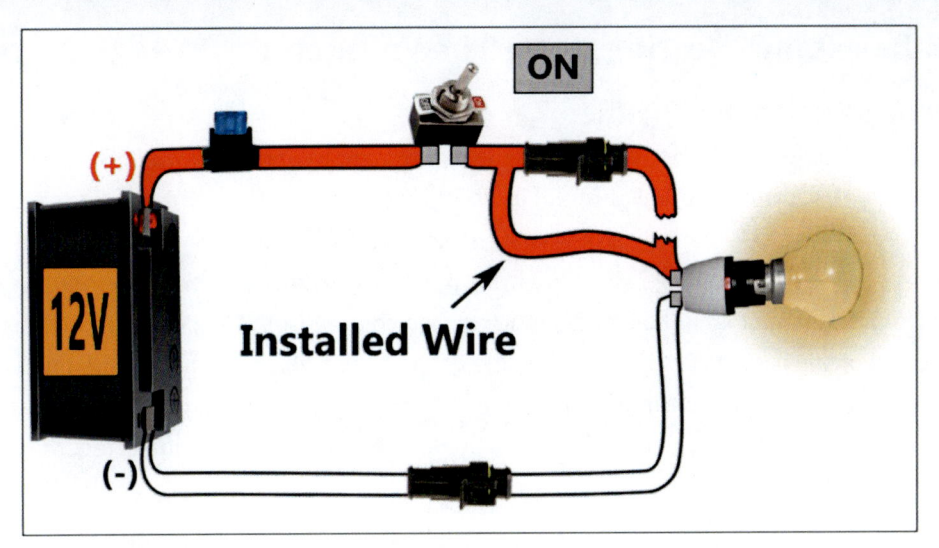

(This is our open circuit with a new wire installed to make it work again. By installing a new wire, we can double confirm the general area of the problem in the circuit. This method can be very useful to pinpoint the area of the open circuit or to even to use as a fix. The installed wire confirmed that the problem was somewhere ahead of the switch going towards the electrical device.)

I do understand that using this method is debatable and some believe it is incorrect and unprofessional to use this as a repair solution. The truth is, sometimes it may be required to resort to this method when it is very difficult to find or gain access to a section of the circuit. The easiest way to fix this problem is to install your own wire directly to the electrical device from a point in the circuit you know there IS voltage present. Better yet, you can disconnect the old wire and run your own new wire to the electrical device.

I know this is not what they teach you in electrical class to do but some people don't have the time to wait or even care how you fix it, they just want it to work and to fix it super fast.

The real trick when installing a new wire is to do it after the fuse and switch. **DO NOT** bypass the fuse or switch because you will not have any circuit protection nor control of when the circuit is ON or OFF, if you do. If I needed to do this, I would do this after the switch or anything that is important such as a fuse or breaker and then make sure my new wire is installed properly and secure. Then if not already insulated make sure that the ends of the new wire are insulated with electrician's tape for protection. If you know how to solder or install a new wire another way, I suggest you make your life easier and give this a try next time you have a difficult open circuit.

The purpose of this chapter was to show you how voltage moves, how to predict and test where an open will be and give you some other neat tips along the way. I hope this is enough to make you understand these key concepts of voltage. In the next chapter we will tackle resistance problems using the voltmeter.

Ch.4: Resistance Testing Using The Voltmeter

Signs of a Resistance Problem.

Whenever a problem with too much resistance in a circuit happens, you will usually see the symptoms in the form of a dim light bulb, a slowly operating motor, a weak heater and other low performances of the electrical device in the circuit. The extra unwanted resistance in the circuit robs some of the voltage available from the power source and prevents the electrical device from working at its full potential. The voltage coming from the power source is shared between all of the resistances that exist inside of the circuit. Because of this, the unwanted resistance will not allow the full power source voltage to reach the device in the circuit. The extra resistance also results in less electricity flowing in the circuit as a side effect of the increase in resistance.

In the following we will list examples of some common problems that are usually the cause of too much resistance in a circuit.

- An electrical wire with damaged, burnt, or missing strands.
- Corroded or damaged electrical connectors.
- Corroded or damaged electrical contacts and terminals.
- Any loose or bad electrical connections.
- A worn-out switch.
- The electrical device itself being worn or having too much resistance internally.

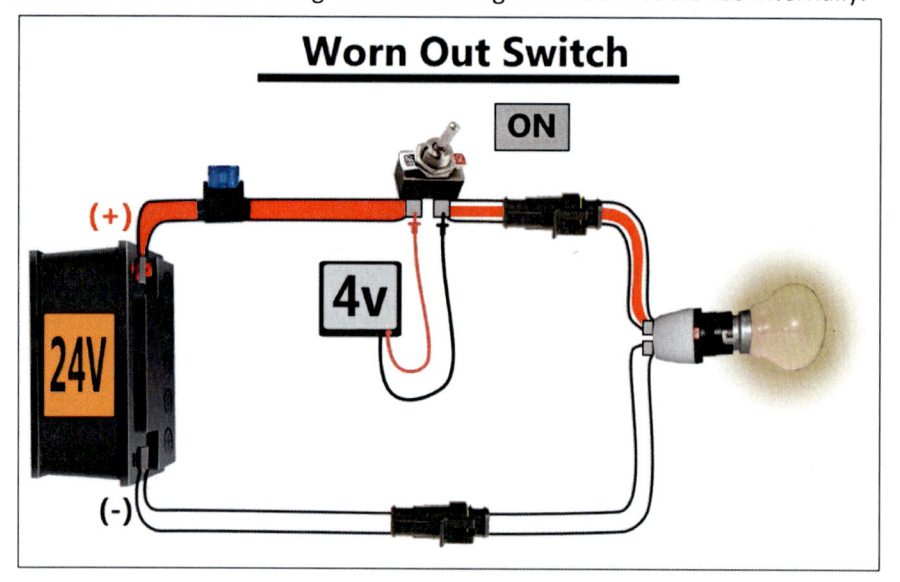

(Here is an example of a 24 volt circuit with an unwanted resistance problem. This circuit has a worn out switch. When this happens the switch doesn't make good contact internally and prevents some of the voltage from one side from passing onto the other side of the switch. It acts as a resistance to electricity and the switch drops some of the voltage that should have been available to the light bulb. The result is a dimly working light bulb and some voltage being lost across the switch. The 4 volt drop across the switch confirms that the switch is worn.)

So where do we start and how can we confirm that we have a resistance problem? Well simple. We do the same first steps as we would any other problem. We begin by checking for voltage at the electrical device of the circuit. In the following we will show another example of a resistance for learning purposes.

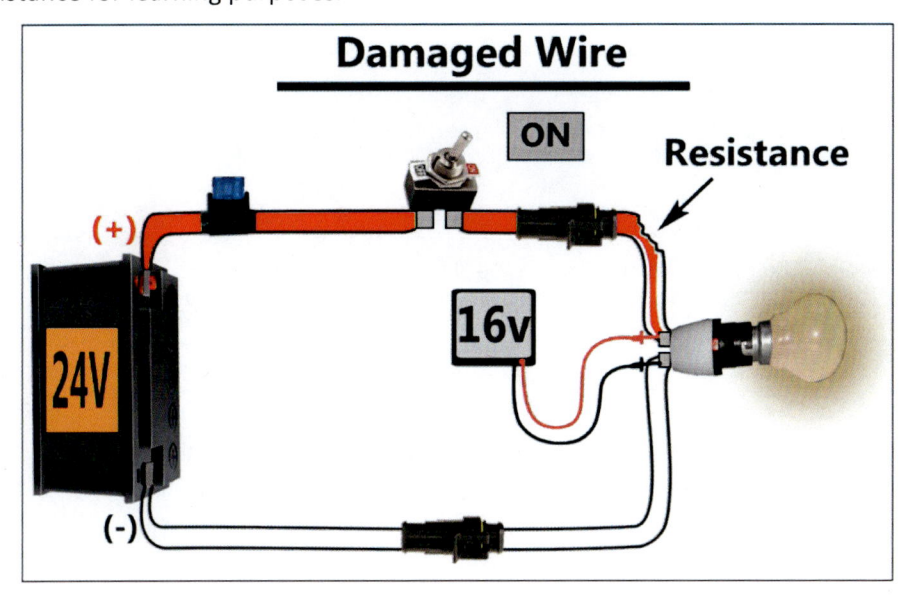

(This is a circuit with a resistance problem in the wiring. You start your testing at the electrical device just as before. A voltage reading that is significantly less than the power source voltage is a good sign of a resistance problem.)

You should generally have the most of the power source's voltage available at the light bulb, this case the 24v. Although as you can see in the illustration, we do not have the 24v on the voltmeter but instead only 16v. This confirms that we do in fact have a resistance problem in this circuit.

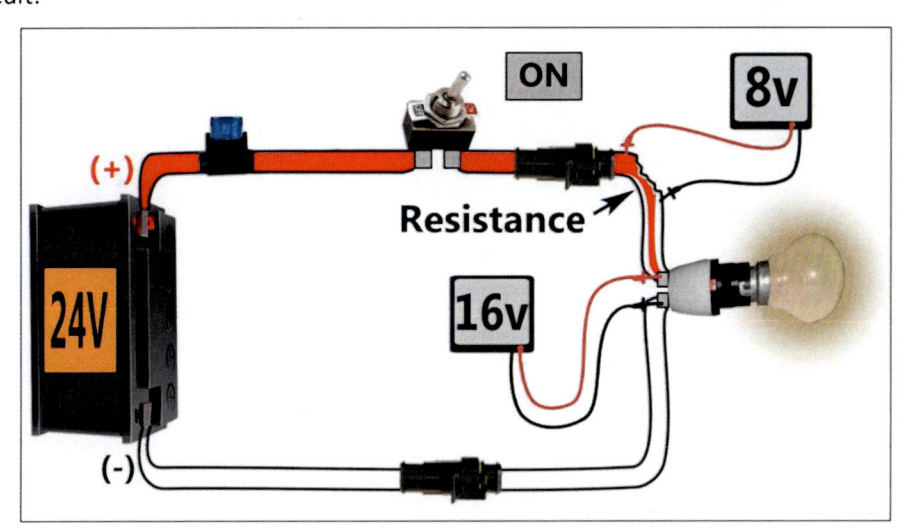

(In this illustration, we have shown the previous circuit but now with another voltmeter installed across the resistance in the circuit. The voltmeter reads 8v across the damaged wire which is exactly what is missing at the light bulb.)

This is made to show you that any unwanted resistances that exists in the circuit WILL have the missing voltage across it. Whether it's a worn switch, a damaged wire, a loose connection, etc. the excess resistance will have voltage being dropped across it that you can measure with your voltmeter.

Reminder: Any tests you do in a circuit with a voltmeter should be done with the circuit ON. Resistance will not drop ANY voltage if the circuit is not ON and electricity isn't flowing. Always test with the circuit turned on and remember to use your safety gear if working with high voltages.

You now know how to tell whether you have a resistance problem by voltage testing at the electrical device first. The trick now is to be able find where exactly the resistance is in the circuit.

The methods of finding the resistance are fairly similar to as if you were looking for an "open" circuit. There are two ways you can approach this. You can either leave one probe to negative and use the other test probe to check the circuit in other areas searching for the missing voltage in the process OR you can start your voltage testing by testing across easily accessible parts of the circuit to see if you can find the resistance faster that way. The choice is yours. Here's what I mean..

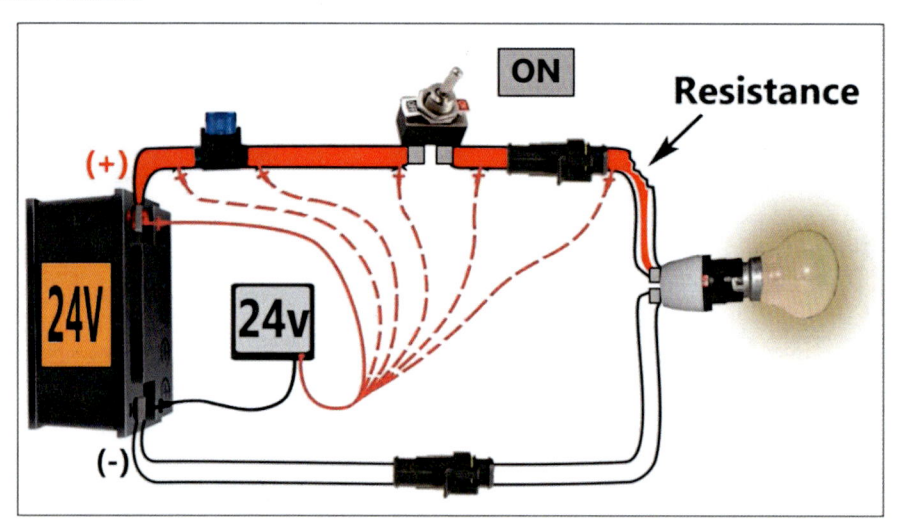

(In this image you can see that we took the first method of testing. One probe is place to negative and the other was used to probe the various parts of the circuit. Before the fuse, after the fuse, before the switch, after the switch, etc.)

Anything involved in the circuit has a potential to become worn or loose and create resistance. Everything up to the point after the power side connector gave a reading of 24v meaning that there is no problem with anything in the circuit up to that point. Let's see what happens when we are nearing the resistance.

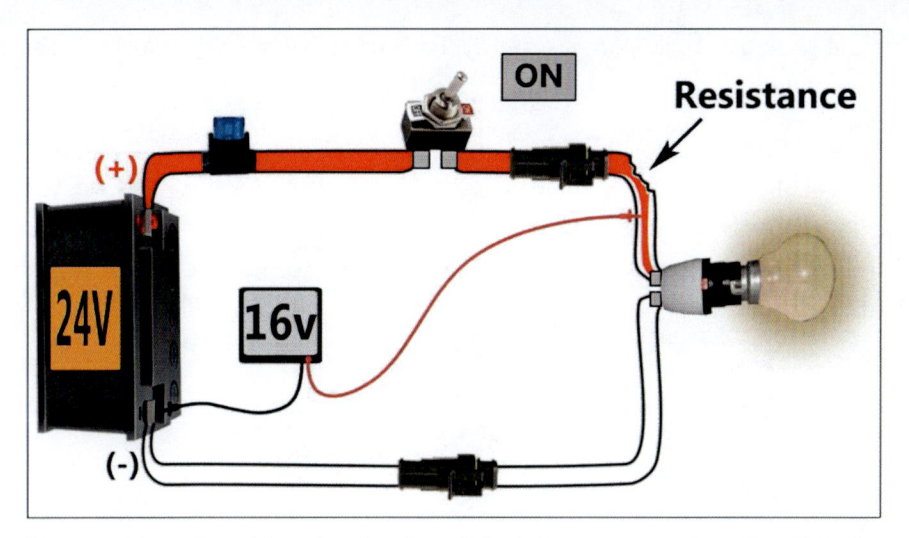

(Here we have continued probing the circuit and find that at one point after the power side connector we read 16v. This is the same voltage that was available at our light bulb.)

This voltage reading of 16v means that from this point on in the circuit all the way up to the light bulb nothing else is dropping voltage. It also means that we just passed the unwanted resistance and we are very near to it. All that is left now is to check the circuit from the point where your red probe was before to the point where it is now.

By following this simple process, you should be able to track down any resistance problem whether it is a damaged wire, a connector, terminals and contacts, a worn switch or any other worn part in the circuit. Repair or replace the damaged parts that are found as necessary.

Let's see an example of resistance in the negative side now and how it slightly differs.

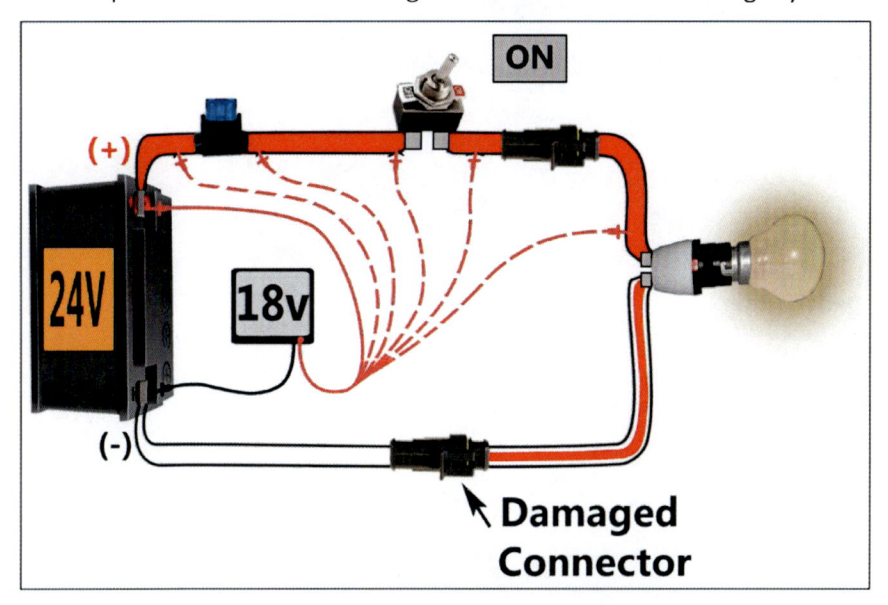

(Here we have another circuit now with a resistance problem in the negative side. We have done just like we did before, looking for a voltage drop across the power side caused by the resistance in the circuit. Throughout the whole power side, we noticed that we kept measuring the same 18v that we measured at the light bulb when we first confirmed a resistance problem.)

There are two possibilities to what is going on. One, maybe the voltage is being dropped right in the beginning of the power side or at the positive battery connection. Or two, the resistance is in the negative side of the circuit. Let's just say the battery connections were checked and looked good, now the only other possibility is to start testing the negative side of the circuit.

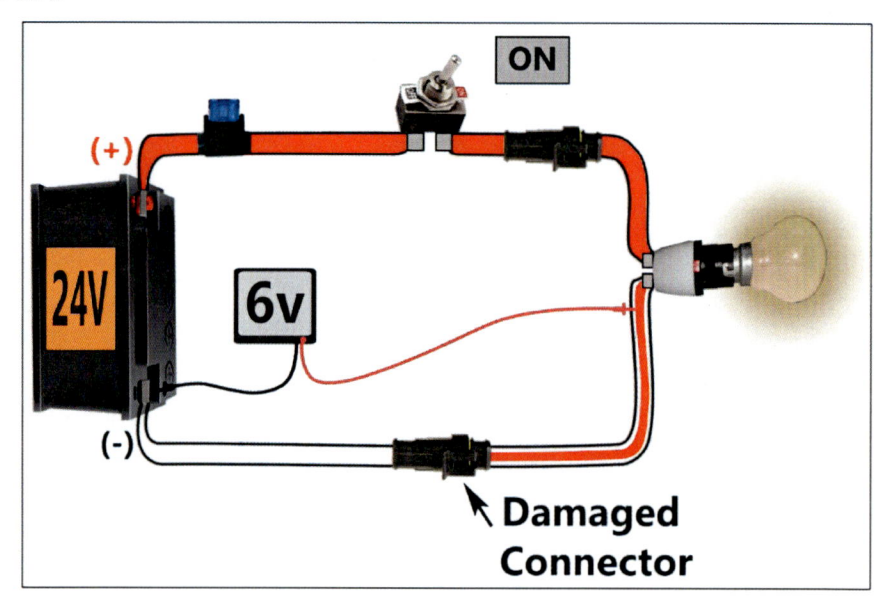

(In this illustration, we have moved to now testing the other side of the electrical device, the negative side of the circuit. Right away we finally see a different voltage reading. We measured 6v in the negative side of the bulb.)

This indicates that we are on the right track towards finding the resistance problem. Don't stop now continue testing!

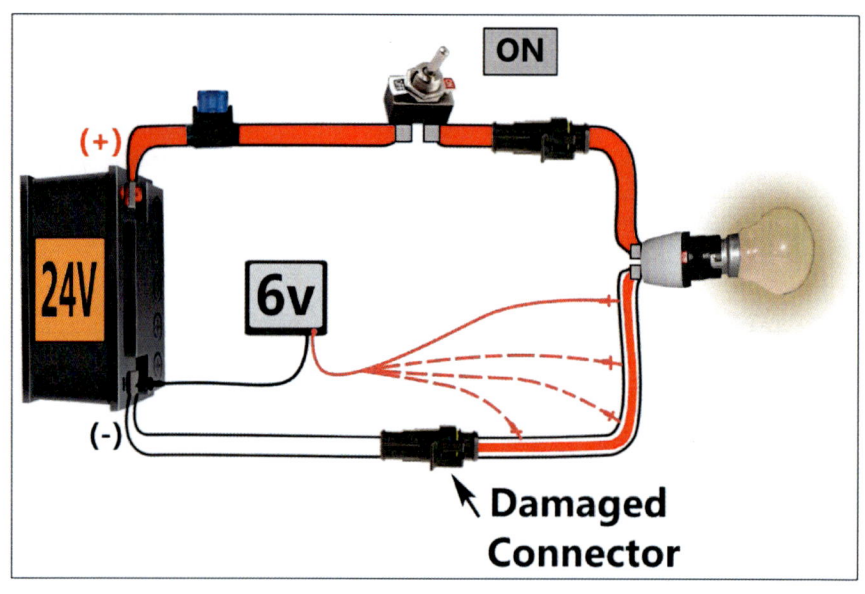

(Here we continue testing on the negative side's wiring all the way up to a negative side connector. So far the voltmeter still continues to read 6v. Let's test the other side of the connector now.)

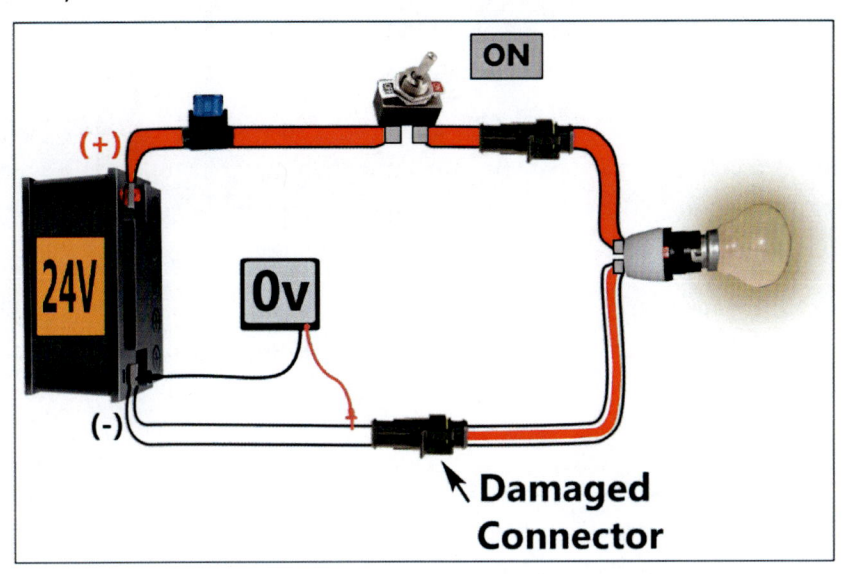

(Here we have moved on to probing the other side of the connector and we noticed that the reading dropped to 0v. This means that our problem is somewhere between where the probe was before and where the probe is now.)

The only thing between those areas is the connector, so we go to disconnect it and right away we see that the inside of it is completely damaged. We have found the problem. Now all that is left is to repair or replace the connector.

We have now seen what it takes to find a resistance problem with the voltmeter but as a side note this last problem could have been very time consuming since we basically went through almost the entire circuit. Let's see the second method we could have used for voltage testing that might have found the problem faster.

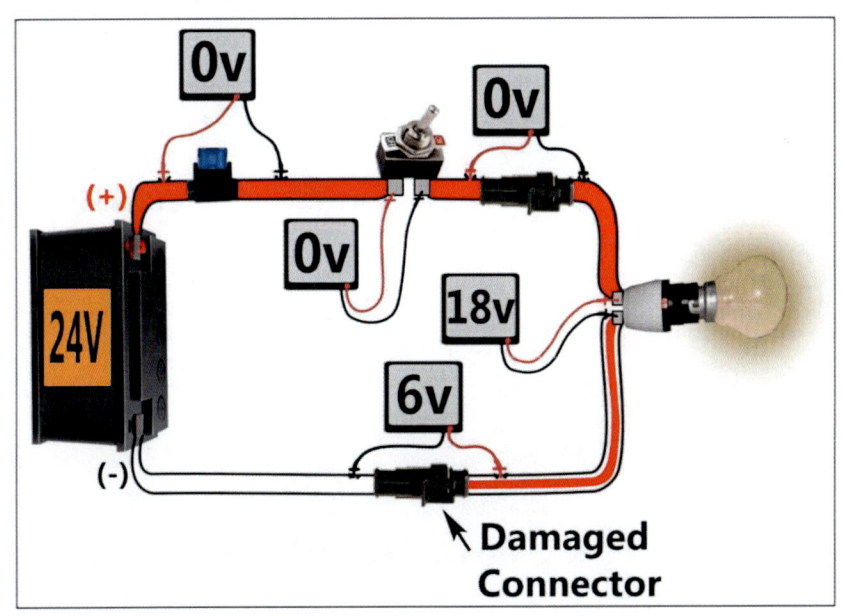

40

(What if instead of putting one probe to negative and using the other probe to test different areas, like we did before, we instead went directly after the parts of the circuit. If we checked the voltage drop across the switch, the fuse, the power source connections and the various connectors of the circuit first, we may have found the problem sooner. We would find that the connector on the negative side had a voltage drop of 6v and that this connector is the cause of the resistance in the circuit.)

I have added extra voltmeters so that you understand what is going on and you see what the different areas would read with this problem. The voltmeter in the bottom of the circuit found the missing voltage at the negative side connector.

Neither one of these test methods is better than the other as you will use both methods in real life. It's simply a matter of which one is easier to do depending on the location of the circuit and what is easier to access. These methods can be applied to both resistance and open circuits.

Note About Replacing Parts That Have Tested Bad: Before replacing any suspected bad component such as a switch or electrical device, etc. Remember that before you spend money replacing it, you should check how well the terminals and contacts were holding on to the bad part. If there was a good and tight connection when you found it then its ok to just replace the part. But if not, you may find that the connections to the part could have just been loose and that the resistance problem was caused by a simple loose connection. Sometimes I have even found that removing and reinstalling the "bad part" seemed to fix the problem by obtaining a better connection.

Here is how you know if any connection is good or not. Let's use the battery clamps as an example for testing purposes.

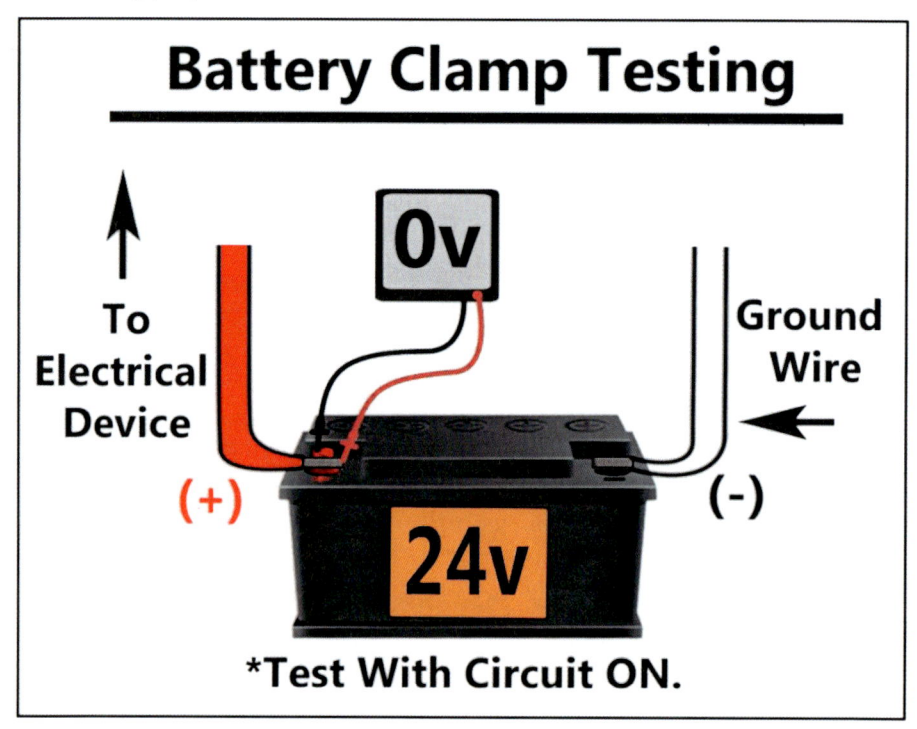

Battery Clamp Testing

To Electrical Device (+)

0v

Ground Wire (-)

24v

*Test With Circuit ON.

(This is an example of voltage testing contacts and terminals. This image shows the battery clamps being tested to see if they are any good. Turn the circuit ON and then place one probe on one side of the connection (the positive battery post) and the other probe on the other part of the connection (the metal clamp). This will tell you how much voltage is being dropped between these two points. The voltage across the connection should be near 0v.)

If there IS a voltage across a terminal or contact or any other electrical connections this means you have a bad connection.

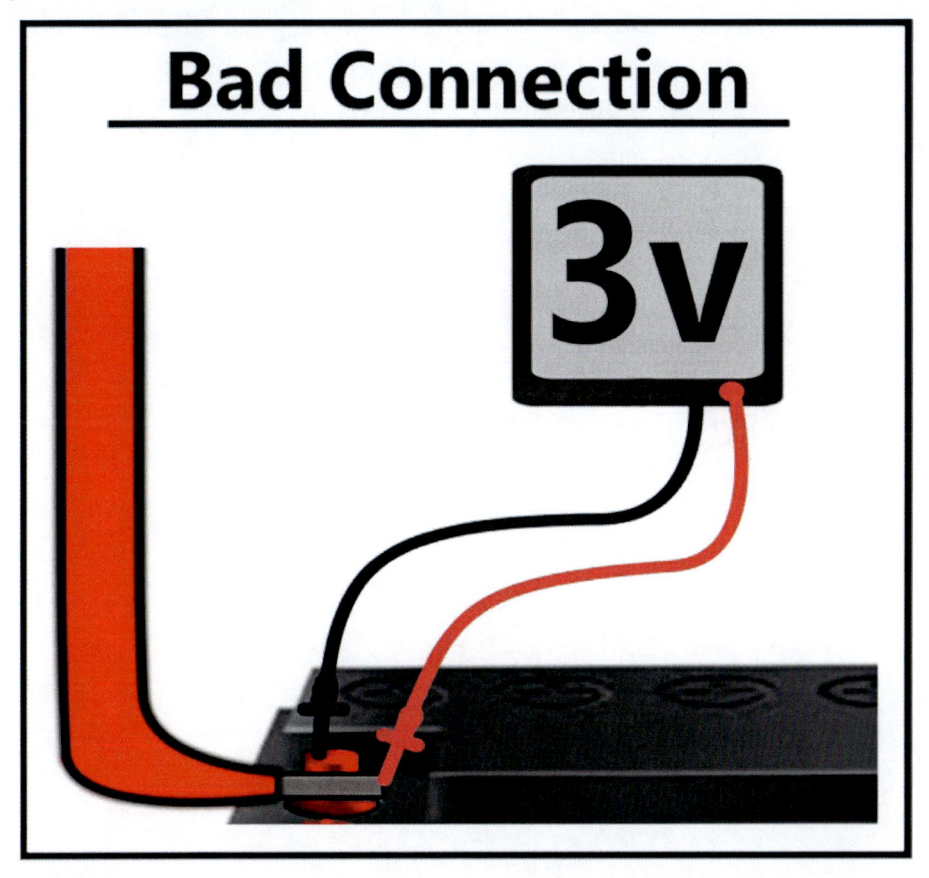

![Bad Connection - 3v]

(This is an example of a voltage reading that confirms a bad connection. It could either be loose, corroded, worn or damaged. Repair or replacement is required as necessary.)

Helpful Reminder: Remember that the electrical device itself can be the one with the excessive resistance. If you have the proper voltage at the electrical device but still see poor performance, make sure that the terminals and contacts on the wiring are secure and that the electrical device is not loosely connected. If the connections are good and still you have a poorly operating device, chances are the device is the one that is bad internally.

A Tip to Save You Time and Frustration on the Job: Tighten and check every connection you see. Wiggle test it and make sure they are on tight. You might end up fixing a problem just with doing this simple check alone. Do not overlook the little things as they are usually the problem.

More Than One Unwanted Resistances: IF you have more than one resistance in the circuit, the voltages of the known and already found resistances will not add up to the battery's voltage. That is how you will know there is still an unwanted resistance somewhere else. After you have cracked down on a resistance problem, make sure to check that the voltage across it PLUS the voltage available at the electrical device adds up be near or the same amount of voltage as the voltage from the power source.

As an example let us say that when we found the bad connector in the 24 volt circuit, we voltage tested it and got a reading of only 2v across it. Then there's also the 18v we tested across the electrical device. This adds up to 20 volts. 18v+2v=20v. The battery's voltage is 24v so this means there is still a resistance somewhere else robbing 4v! If there is ever still a very significant amount of voltage missing even after fixing a resistance problem, then you may have more than one resistance problem in the circuit. Get out your voltmeter and start testing the circuit.

Ch.5: Miscellaneous Tips

Note About Electrical Corrosion:

Wherever you find corrosion in a circuit, there is likely to be an electrical problem. This green gooey stuff happens when the electrical wires or electrical terminals in a circuit are not shielded from moisture. This moisture penetrates exposed wiring and terminals and causes the metal inside that carries the electricity to begin to rot. The green goo is the resulting residue from the rotting metals.

Once a wire or terminal becomes corroded they become brittle or weakened and WILL NOT allow electricity to flow as well as it did before. If you see this anywhere in a circuit, it will likely be the source to a resistance problem or even an open circuit in more extreme cases.

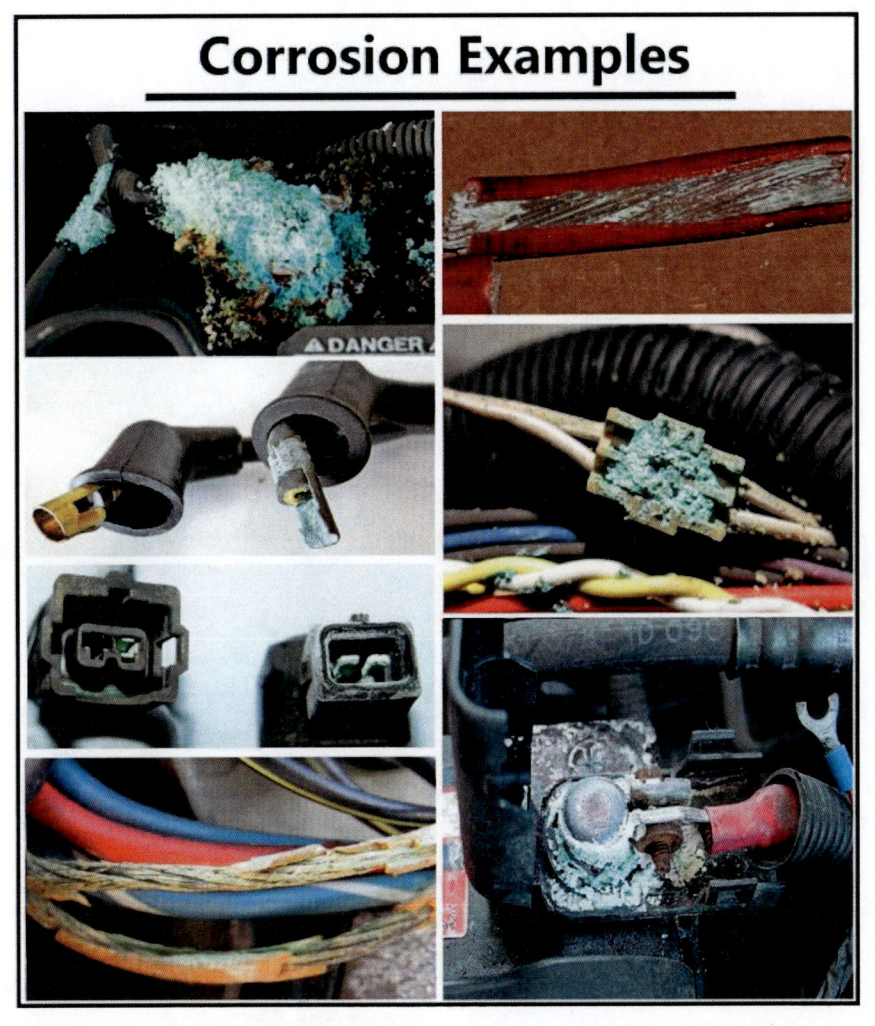

Corrosion Examples

(Example of corrosion on battery cables, wiring, connector terminals and other connections. This sort of corrosion will likely become a resistance problem or open circuit if left untreated. If the corrosion has not already caused too much damage to the conductive metal parts of the circuit, it may be able to be fixed by cleaning it away using electrical parts cleaner and then reinsulating the area to protect it from further corrosion damage. If the damage is too great, then the corroded part must be replaced.)

Also, if a part of a wire shows signs of corrosion it is likely that a larger section has already began to corrode as well. The only way to fix a corroded wire is to replace the bad section of the wire with a new piece or run a new wire.

Using Contacts As Test Points:

Using Contacts As Test Points

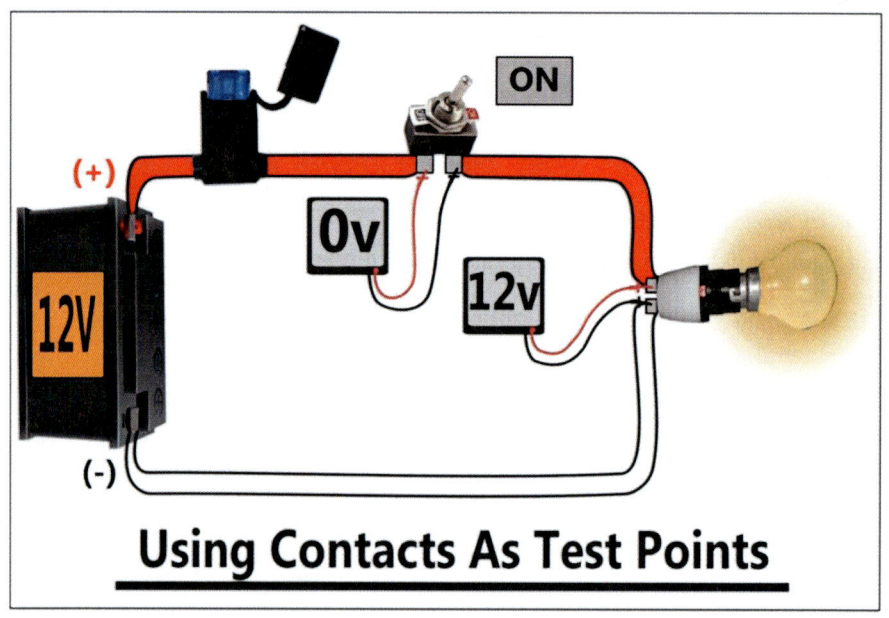

(You can use contacts as test points for voltage measurement if it's easier to access. If readily available, there is no need to pierce any wire for a voltage measurement. Remember that you must have the circuit ON for proper voltage testing.)

Using T-pins or Sewing Needles:

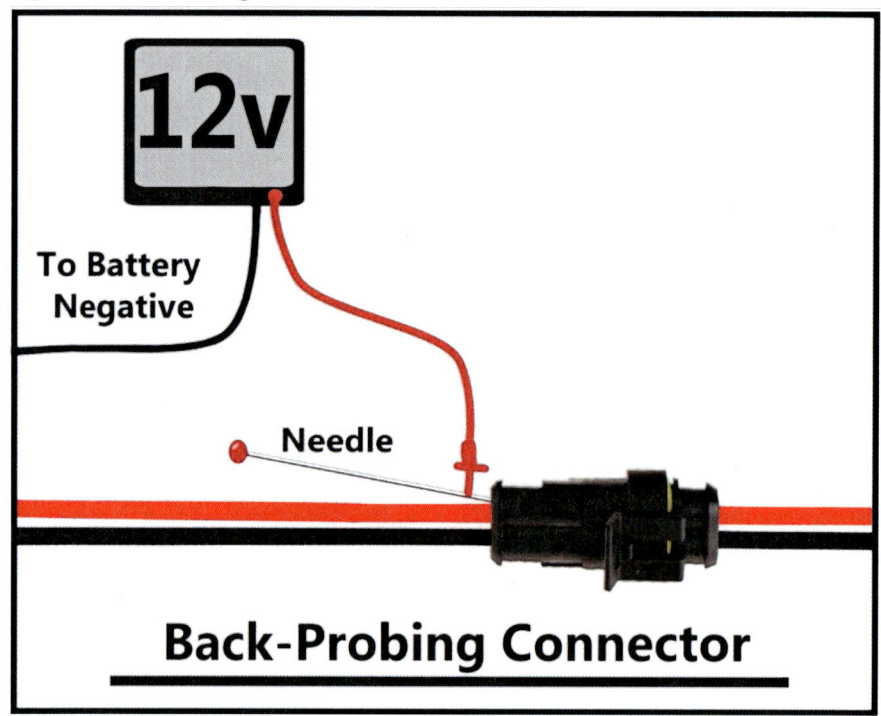

Back-Probing Connector

(For those electricians who don't like to poke wires then this is the method for you. Testing at connectors can be done by slipping in a t-pin or sewing needle through the back of the connector to make a test point. The trick to this method is to make sure the tip of the pin is contacting the metal terminal on the inside of the connector. If you don't contact the metal terminal inside the connector you will not read anything when it's time to use it as a test point for voltage measurements. This method takes a little getting used to in order to tell if the t-pin is contacting the inside or not. Try it a few times until you get it right and make sure the circuit is ON while voltage testing.)

Using a Smaller Fuse:

If while on the job you do not have the proper fuse for the replacement of a blown one, and you need to get the circuit to work, you can temporarily substitute a slightly smaller rated fuse in its place. As long as this smaller fuse does not also blow when installed then the circuit should work fine. NEVER substitute a fuse with a larger rating fuse, only substitute by using a slightly smaller rating than the original.

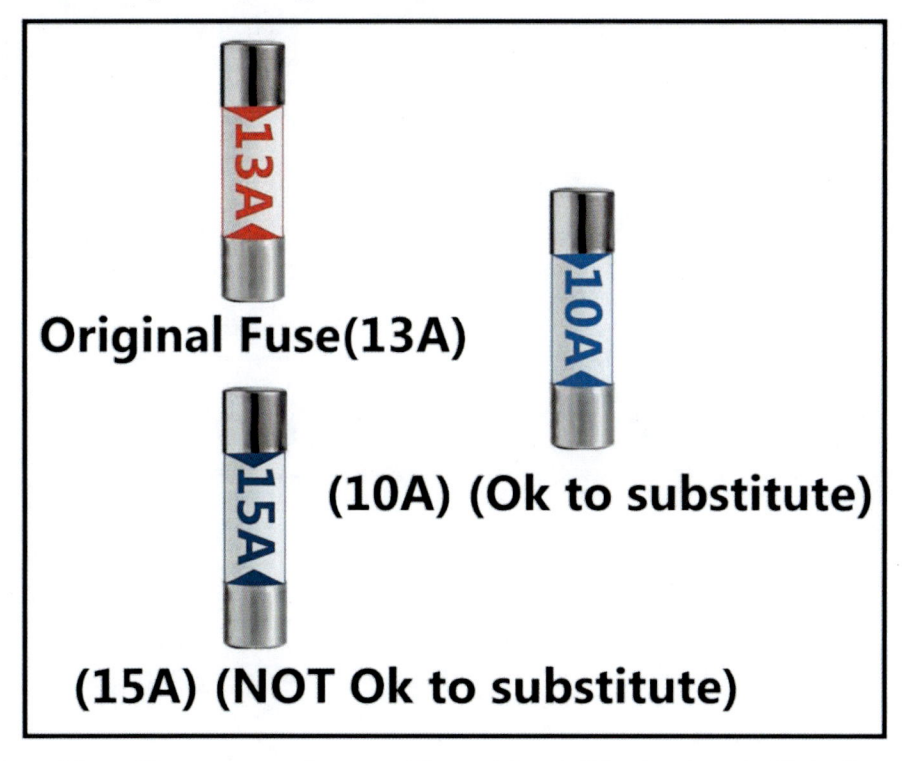

Original Fuse(13A)

(10A) (Ok to substitute)

(15A) (NOT Ok to substitute)

(If you do not have the proper replacement fuse at hand, it is okay to substitute the original fuse with a slightly smaller rating fuse. **DO NOT** replace the original fuse with a large rating fuse because you can cause damage to the circuit if there is a shorting problem.)

How to Solder and Repair a Wire:

If you ever come across a broken or damaged wire, you will know that many times its easier to repair the original wire than to replace the entire thing. It is possible to repair instead and reconnect the broken wires by using soldering.

The tools you will need are a soldering iron, a roll of thin rosin core solder, a helping hand tool and a wet sponge to clean the tip of the soldering iron. Let see exactly how the process of soldering can be done...

(These are all the tools you will need for soldering a wire. A soldering iron, a roll of very thin rosin core solder, a wet sponge to clean the tip of the soldering iron and a mini helping hand to hold everything in place. The brass mesh tip cleaner can also be used as another option for cleaning the soldering tip.)

The following is a step by step process on soldering a wire. Read on for further instructions.

(Step 1: Connect the soldering iron and wait for it to heat up completely (may take up to **15 mins**). While it heats up, wet your sponge to use for tip cleaning the soldering iron. Then once the iron is hot, take some time to steam clean the tip of the soldering iron by wiping the hot tip on a clean wet sponge. Wipe the tip clean on all sides. Then once it is clean, it is ready for soldering. The brass mesh soldering tip cleaner can also be substituted for the wet sponge cleaning. Do not skip this step otherwise your iron may not be able to transfer enough heat to solder properly.)

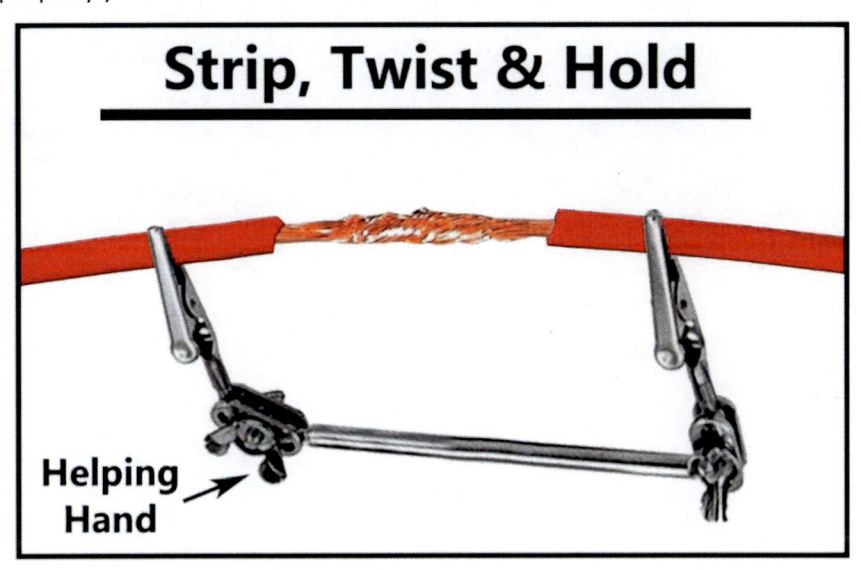

Strip, Twist & Hold

Helping Hand

(Step 2: Get the wires that you are to solder and strip the ends off revealing the copper strands of the wire. Get the two copper ends and twist them onto each other. Hold them from unraveling the twist by clamping on a helping hand. I call this method the Strip, Twist and Hold. There are other methods for this step of course, but just follow whatever your comfortable with to hold the wires in place for the soldering process.)

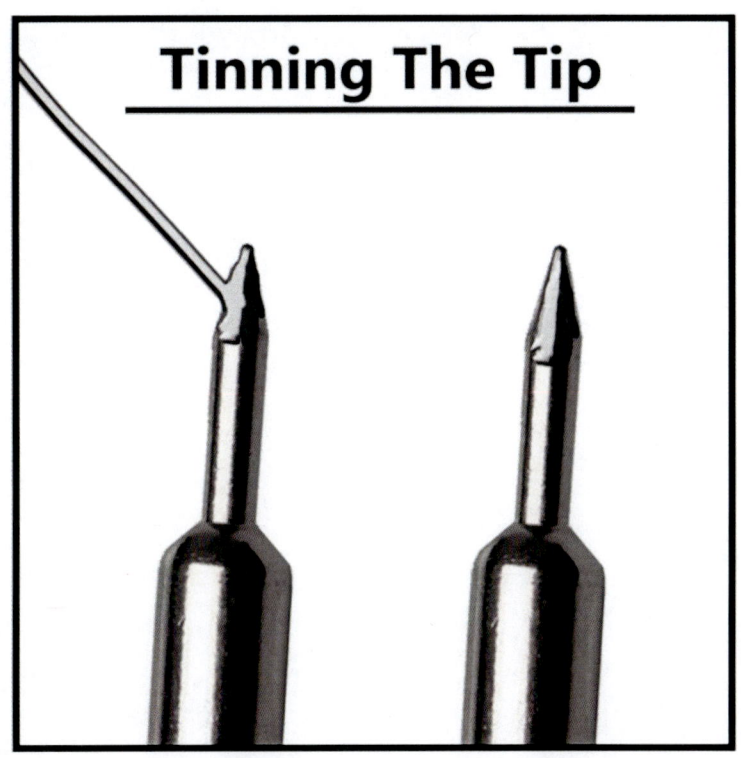

Tinning The Tip

(Step 3: Melt a small amount of solder onto your soldering iron tip before soldering the wire. I recommend using the thin sized rosin core solder because it is easier to melt. This is step is known as "tinning the tip".)

Touch The Tip To The Wire, Add Solder To The Wire.

(Step 4: Touch the wire with the tinned tip of the soldering iron to begin to transfer the heat needed for soldering. Once the wire has become hot enough, add solder to the wire itself close to the iron tip. The solder should melt easily and flow onto the wire just by continuing to add solder to the wire. Once you have complete coverage of solder onto the wire, remove the the soldering iron and allow the wire to cool. Do not pull or stress the wire while it cools down.)

The result should be a solid solder connection. Remember that once the iron has heated the wire enough, the solder should easily melt onto the wire and flow just by touching the solder to the hot wire. Once you are satisfied with the coverage and have achieved a shiny solder joint, remove the iron and allow for the wire to cool. By following these easy steps you should have been able to get a solid solder joint that wont come apart once it has cooled.

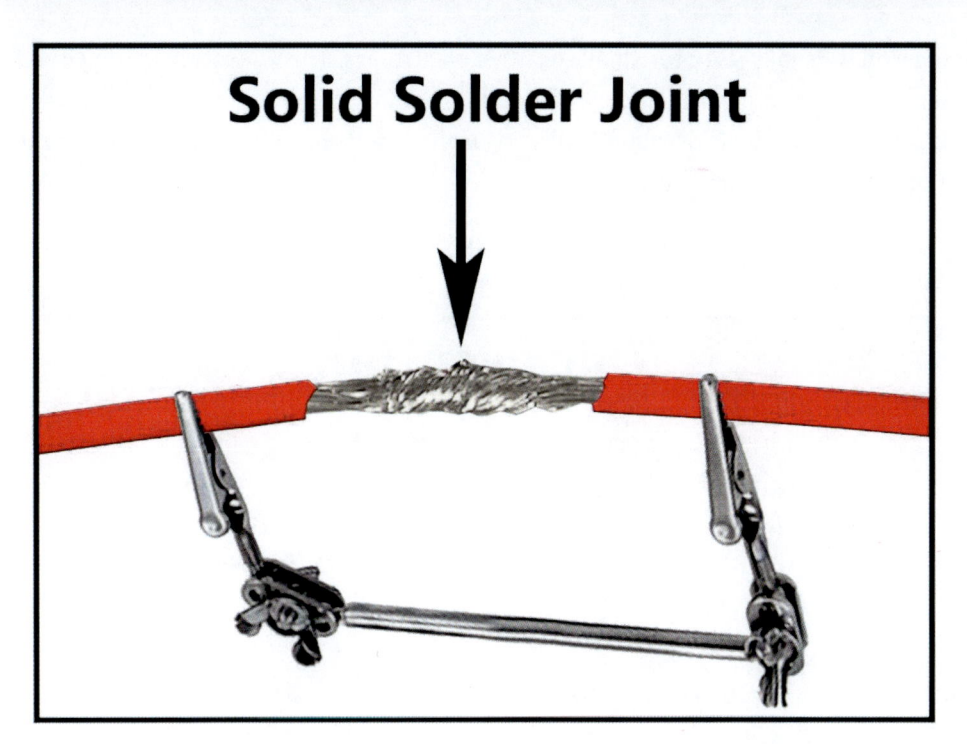

Solid Solder Joint

(This is what a good solder joint should look like. It should be shiny, smooth and strong and should not come apart when pulled on once it has cooled.)

Common Soldering Problems:

Soldering can seem like a tough process to learn for the beginner without guidance of what they are doing wrong. If you didn't get it right the first time don't worry. Here are some common reasons as to why the soldering process doesnt go as planned.

(Reason #1): Not enough heat from the soldering iron. May need a higher watt soldering iron for the task at hand.

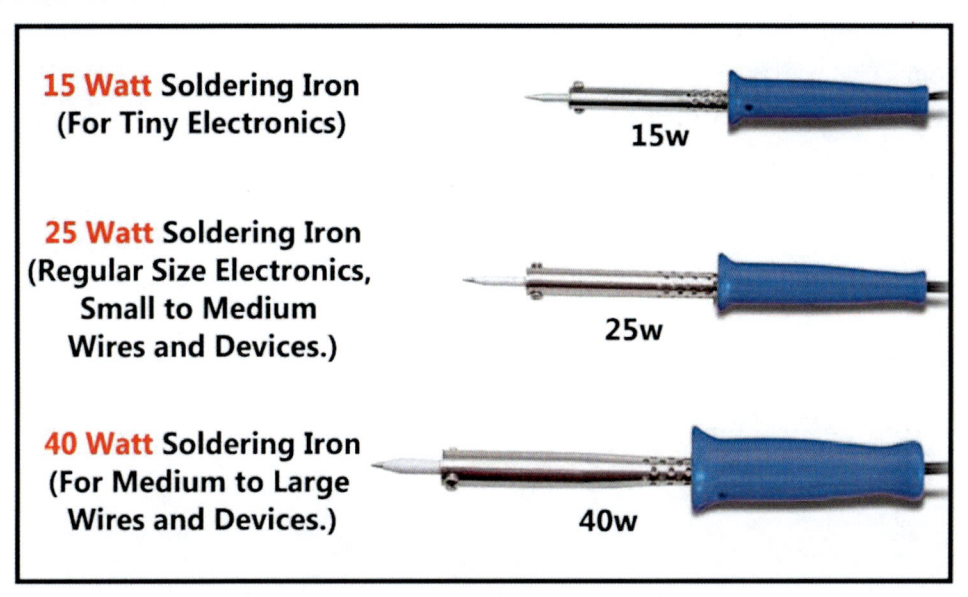

15 Watt Soldering Iron
(For Tiny Electronics)

25 Watt Soldering Iron
(Regular Size Electronics, Small to Medium Wires and Devices.)

40 Watt Soldering Iron
(For Medium to Large Wires and Devices.)

(There are various sizes and wattages that soldering irons are rated at. Make sure the soldering iron you purchase is strong enough for the job you intend to use it for. The 15 watt soldering iron is usually used for small sensitive electronics. The 25 watt soldering iron is a general purpose soldering iron and can normally accomplish the small to medium size soldering jobs. The 40 watt or higher soldering iron is used when there is a need for extra heat during medium to large soldering jobs.)

(Reason #2): The solder being used is too thick for the application and isnt melting easily on the hot iron tip.

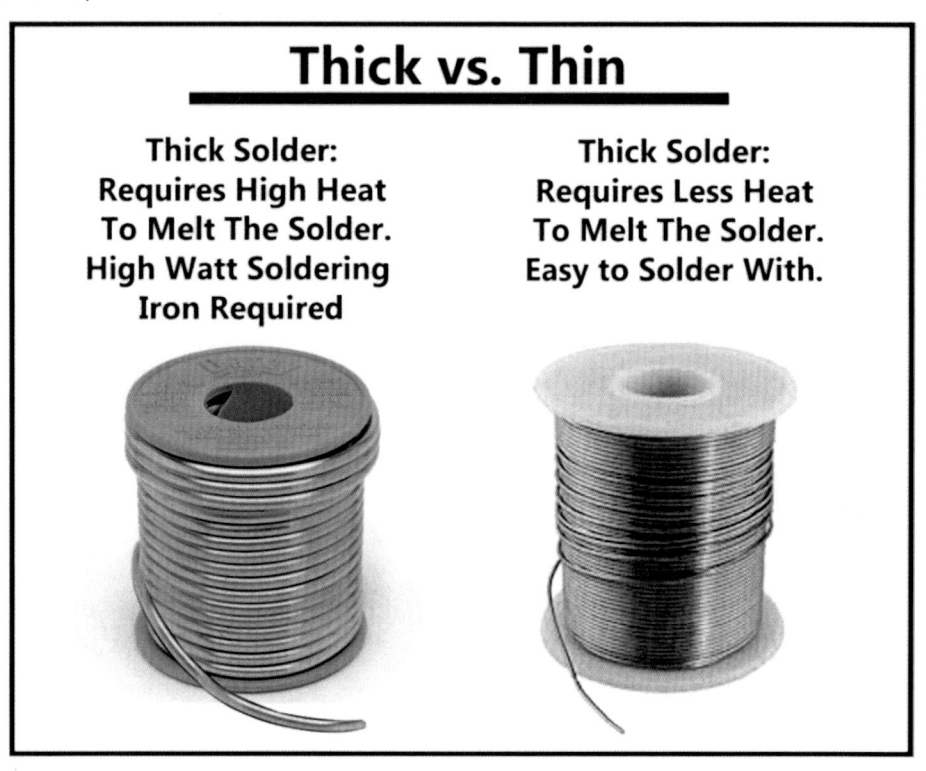

Thick vs. Thin

Thick Solder:
Requires High Heat
To Melt The Solder.
High Watt Soldering
Iron Required

Thick Solder:
Requires Less Heat
To Melt The Solder.
Easy to Solder With.

(The size of the solder that you are using is very important to the success of the soldering process. The thick solder is usually harder and slower to work with because it requires a lot of heat from a high wattage soldering iron used for heavier duty applications. The thin solder is easier to use because it will melt away faster when exposed to the heat of the soldering iron. The goal is to use a solder thickness recommended for your soldering iron size. If all else fails go with a thinner size solder.)

(Reason #3): The tip of the soldering iron is dirty or worn out and isnt able to transfer heat properly.

Cleaned vs. Dirty

(In order to allow for proper heat transfer from the hot soldering tip to the wire or the part to be soldered, the tip must always be clean. This can be achieved by cleaning the tip before use either with a wet sponge or a brass mesh solder tip cleaner.)

(Reason #4): Wrong size/type soldering tip being used.

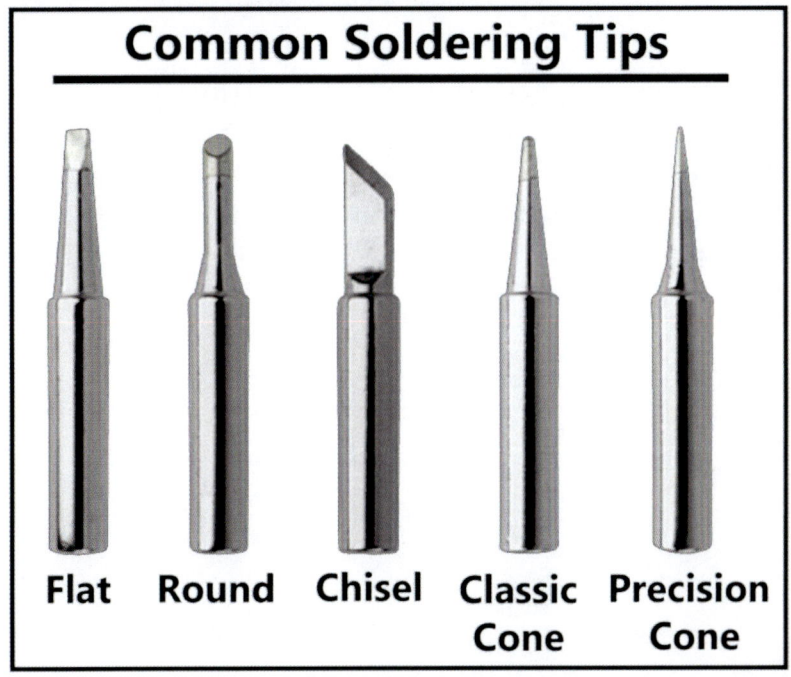

Common Soldering Tips

Flat   Round   Chisel   Classic Cone   Precision Cone

(There are various types of soldering iron tips available, each with its own unique purpose. The flat and round soldering tips are used for when a larger contact area is desired for transferring heat faster to the part to be soldered. The round tip allows for a larger contact area at an angle when compared to the flat tip. The chisel tip is normally used for cutting soft materials or engraving. The conical style tips are used for when a smaller contact area and slower heat transfer is desired or when soldering a very small piece that may burn if exposed to excessive heat.)

(Reason #5): Dirty, burnt, old or oxidized wires

Burnt, Dirty or Corroded Wiring

(One other problem that is usually overlooked and may be preventing you from soldering is oxidation, dirt or overheated sections on the piece to be soldered. The wires or piece to be soldered should always be thoroughly cleaned before soldering otherwise the solder will not flow properly. Usually if a wire is not badly damaged then a wire brush along with some electrical parts cleaner or baking soda solution can restore the wires. If the wire to be soldered is found to be heavily burnt, damaged or oxidized like in the images shown, then the entire wire should instead be replaced.)

Don't be too discouraged about not being able to solder properly the first time just try it again and follow the guidelines if there is any trouble. One other tip is to use solder flux to make the solder flow a lot faster.

Solder Flux

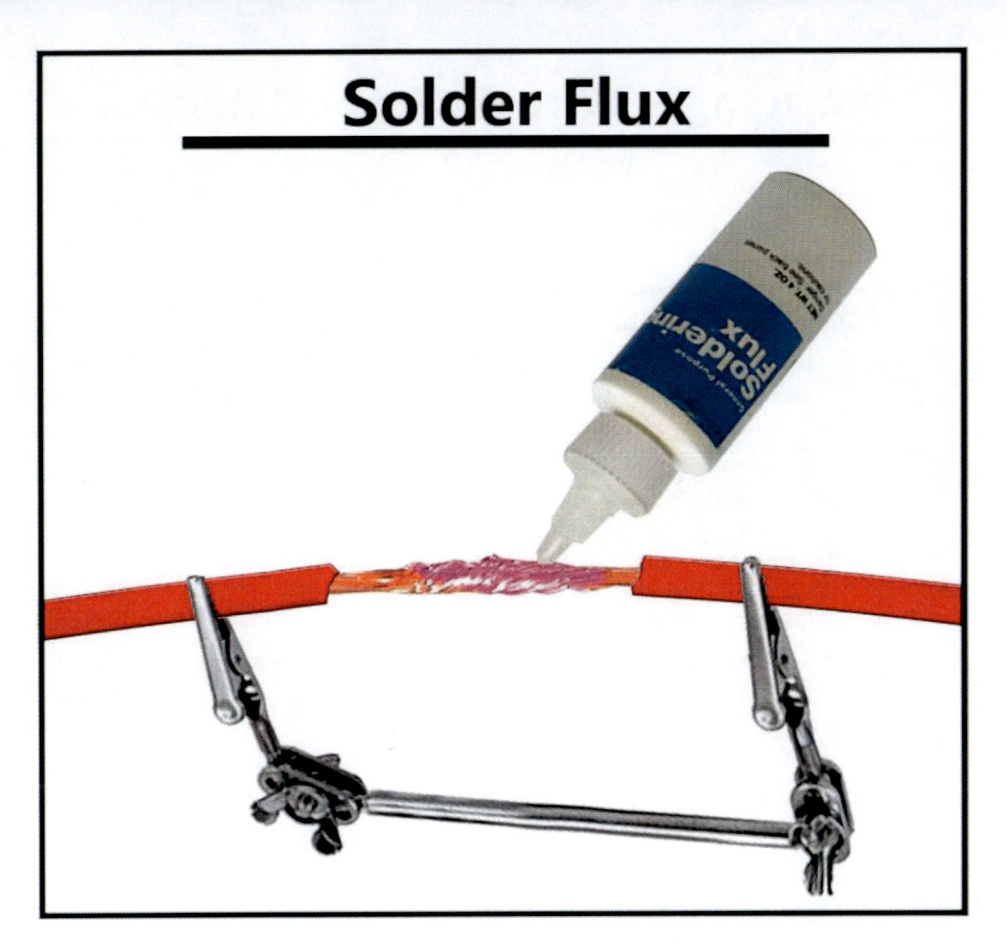

(Example of liquid soldering flux being used on the wire before the soldering process begins. This allows the solder to flow better throughout the wire when adding the solder to the heated wire.)

Final Tips for Soldering:
There are two most commonly seen types of solder cores available at your everyday electronics or hardware store. There is the Acid Core and the Rosin Core type solder. Rosin core is a non-corrosive type solder. The mild nature of the rosin core limits it to be used in only copper and brass soldering applications. The acid core is a more aggressive type designed to solder steel as well as other hardened metals. The acid core should only be reserved for use in steels and iron. The acid core will degrade the copper in normal wires and the soldering tip as well if used. The best option for copper wire and other normal soldering applications is the rosin core solder.

Probe Extensions
Now let me show you a neat trick about voltage being present where the open is. Let's check a battery's voltage by connecting very long wires to each post and testing it from a very remote area…

Using Jumper Wire As Probe Extensions

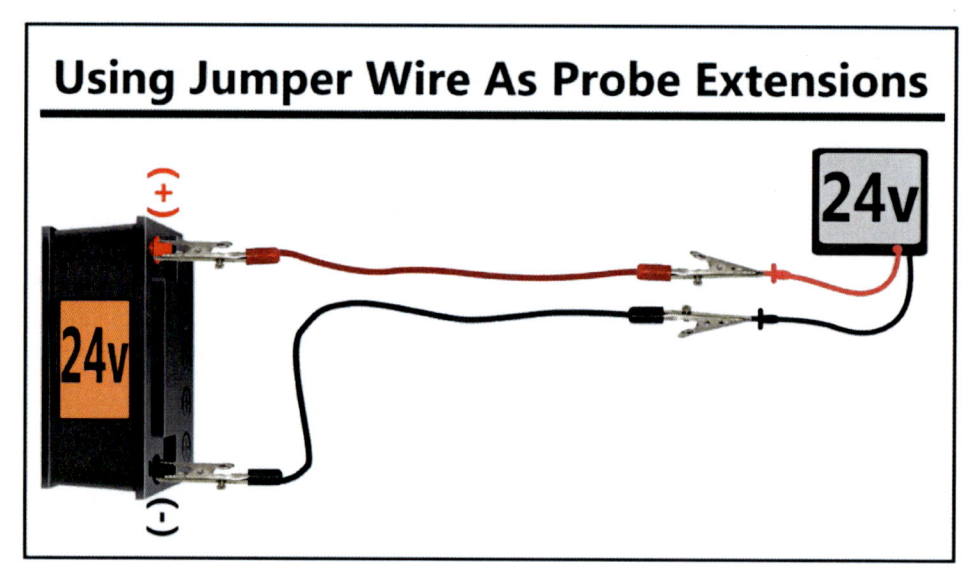

(This image shows a battery with very long wires attached to each post. The voltmeter is reading the battery's voltage from a long distance away. If the wires were long enough you could test your car's battery voltage from your couch while you watched TV!!)

Sometimes your probes may not be long enough on their own and may require a jumper extension to reach to desire probing area. This is one easy method of extending your meter probes.

Helpful Tips:

Now a tip for your meter's probe tips... Sometimes during testing you may be required to accomplish your electrical testing while at the same time doing something else with your hands. If you only have the regular needle style probes that come with your meter you may have to improvise to finished the task. Many times a more hands-free method of testing will be needed.

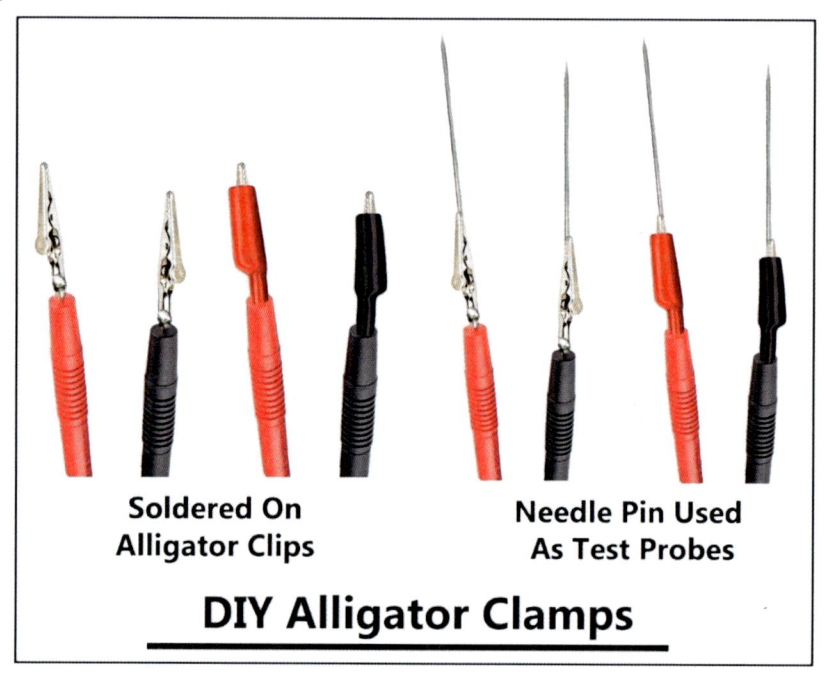

Soldered On Alligator Clips

Needle Pin Used As Test Probes

DIY Alligator Clamps

(Quite a few meter probes that I own have alligator clips soldered onto the tips. This is easily done by bending the ends of the alligator clamps to fit tightly on your needle probes. Then just add solder to the area to weld the two pieces together and now you have your very own clamp probes.)

There are add-on alligator clamps that you can buy specifically designed for your meter's probe tips. These add on clamps that are for sale don't require you to solder and are designed to be easily installed or removed when needed. In fact, there are many add-on tools for your meter probes to allow you to do a variety of neat things including hands free testing.

I usually get by with soldering on very inexpensive alligator clips to my probe's tips. From this I can use the alligator clips as a regular tip, clamp on to a terminal or contact and even hold a very thin needle or T-pin to use as an ultra-long thin probe tip. The image to the right shows an example of a pin being held inside the jaws of the alligator clips. These come in very handy when trying to access places that require a long yet thin tip to make contact for a voltage test.

A Word Of Caution: Although it is always nice to save money by making your own test probe you must also be aware of the potential dangers. If you decide to make the DIY test probes similar to the ones shown here, just be very careful not to use these on circuits over 48v without the clips being fully insulated. If you are working on a higher voltage it is always best to use professionally insulated add on tips for your meter that are designed to safely handle up to certain amount of voltage.

How To Tell If Amps are Flowing Using Your Voltmeter.

When you find a voltage drop across an electrical part during testing, you know for a fact that the circuit is on and electricity IS flowing. There can be **NO** voltage drop without amperage or electricity flowing inside a circuit. The circuit must be ON for any voltage drops to occur. This is actually how you can use your voltmeter to tell if there are amps flowing inside a circuit.

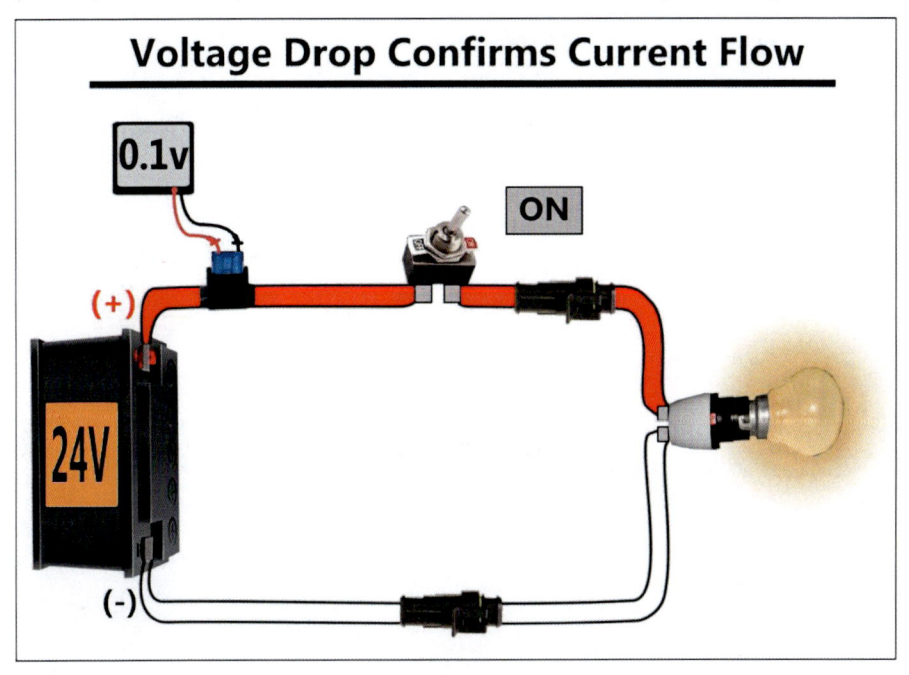

(This illustration shows an example of using the voltmeter to see if the circuit is on and amps are flowing. Testing for a voltage drop across the fuse is usually the easiest way to tell if a circuit its ON or not when you cant physically see the rest of the circuit from where you are. In this image you see a reading of 0.1v. Although this is a tiny voltage across the fuse, it tells you that in fact electricity IS flowing through the circuit. This can serve as a very useful test when tracking down a problem where you have a circuit that is shorted and stays ON all the time, draining the battery.)

Note: The previous test using the voltmeter to tell if amps are flowing will only work if you have a good enough meter that has the accuracy and resolution to read the small voltages properly. Invest in yourself and invest in an accurate meter and you will never have to worry about the readings "being off" because of the accuracy of your meter.

The Final Important Note about Resistance in a Circuit and Real Life Circuit Measurements:

Resistance and Temperature: When the temperature in a circuit goes up, the resistance of EVERYTHING in the circuit goes up. When the temperature in a circuit goes cold, the resistance of EVERYTHING in the circuit goes down. The way this resistance change happens in a circuit involves complex physics that you don't really want me to explain. So just remember when testing, that resistance of everything in a circuit will go up if the circuit is exposed to heat and the resistance of everything in the circuit will go down if the circuit is exposed to cold temperatures.

Well why is this even important? Well when testing, you have to remember this important fact. **When a lot of electricity is flowing through a circuit, it creates heat. Heat creates Resistance and Resistance creates voltage losses.** This is why this lesson on the effects of temperature matters. In circuits with a lot of amperage flowing, the heat from the electricity will create larger voltage drops that are actually normal.

If you have a circuit with very low amperage or electricity flowing, then it will not create much heat and it doesn't really make a big difference when taking measurements. But if you have a circuit with a large amperage or large amounts of electricity flowing, then you WILL have to pay attention to this rule when measuring voltage.

When you have a large amperage, let's say 20 amps and up for example, you will have a voltage drop created by the heat of electricity flowing. You will have this voltage drop across connections, switches, terminal, contacts, fuses, etc. This will actually be normal voltage drops and should not be thought of as something wrong with the circuit. Instead of a perfect 0v, you might have 0.05 volt or 0.2 volt drop across a switch. You won't see all the voltage you SHOULD have at the electrical device because of this normal voltage drop due to high amperage flow. Instead of 24v at the electrical device you might read 23.6v or 23v. This isn't so much as problem depending on the amperage that is flowing in the circuit.

You have to be able to tell what **IS** a problem and what **IS NOT** a problem. I will explain this skill and much more tips in my next book "Everything Electrical: How to Test Like a Pro: Part 2."

General Ballpark Guide on Ideal Good Voltage Measurements:

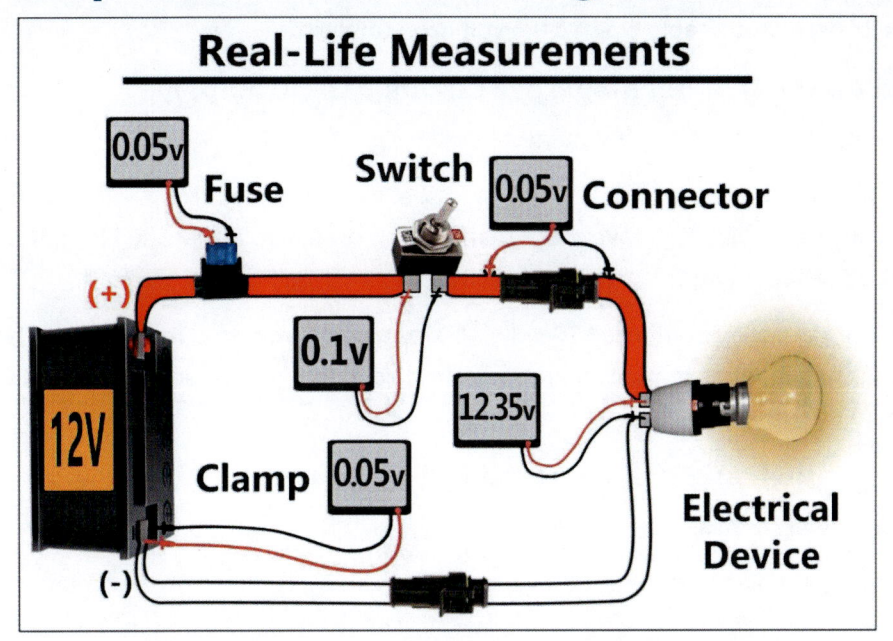

Real-Life Measurements

[Diagram showing a 12V battery circuit with voltage measurements: 0.05v across Fuse, 0.05v across Switch, 0.05v across Connector, 0.1v at one point, 0.05v at Clamp, and 12.35v across the Electrical Device (light bulb).]

When voltage testing you will need to know what a real life good measurement is going to look like. The following guide will serve as only a ballpark guide to what is a "good" measurement. This is NOT an end all guide and there are more specific guides out there that should be followed depending on the electrical applications. With that said, you still need to at least have a general range to know what is "good".

Keep in mind before reading on that there will be variations to the numbers in this guide that will be caused by things like..

- The amperage in the circuit
- The kind of conductor material the electric parts in the circuit are made of
- The length and kind of the electrical wire used in the circuit
- & The accuracy of your testing meter.

The real trick is to understand that the more voltage drop in the circuit or test piece you have, the more problematic it is. Also in a circuit with high amperage, small voltage drops are ACCEPTABLE.

Ideally we would always want to have voltage measurements of 0v across any Switch, Fuse, Wire, Connector, Terminal, Contact or Any Other Electrical Connections. The fact is, depending on the amount of electricity flowing, the voltage across these things will increase due to heat. With this in mind, there has to be at least a ballpark range for you to use to know if a voltage drop is normal or not.

Low Power Low Amperage Circuits (1-10 amps):

Near 0 volts drop. Acceptable Up to a 0.5 volt drop total in the entire circuit.

Medium Power Medium Amperage Circuits (10-30 amps):

Minus 5% of the Power Source's voltage.

Example 1: If the power source is 12.6 volts, - 5% equals a 0.6 volt drop maximum in the entire circuit. You should not have more than 0.6 volt drop across any one place. If there IS voltage drop, it should be spread out throughout the circuit..

Example 2: If the power source is 120 volts, - 5% equals a 6 volt drop maximum in the entire circuit. You should not have more than 6 volt drop across any one place. Again, this voltage drop should be spread out through the circuit..

High Power High Amperage Circuits (30-100amps):

Minus 8% of the Power Source voltage.

Example 1: If the power source is 12.6 volts, - 8% equals a 1.1 volt drop maximum in the entire circuit.

Example 2: If the power source is 120 volts, - 8% equals a 9.6 volt drop maximum in the entire circuit.

Very High Power High Amperage Circuits: (100-250amps):

Minus 12% of the Power Source voltage.

Example 1: If power source is 12.6 volts, - 12% equals a 1.6 volt drop maximum in the entire circuit.

Example 2: If power source is 120 volts, - 12% equals a 14.4 volt drop maximum in the entire circuit.

Remember that you do not want to have all the voltage dropping across any one place. The voltage drop will be spread out throughout the circuit if its normal voltage drop. No switch or connections should have a large amount of the total voltage drop in the circuit across it.

Conclusion To Part 1:

Congradulations on making it this far in the Part 1 of this series. In Part 2 of the series I will pick up where I left off and showing you exactly how you can solve **Intermittent** or randomly occuring electrical problems. Stay tuned on "How To Test Like a Pro".

End Credits:

Thank You For Reading And Good Luck! I Hope That This Book Had At The Very Least One Thing That You Can Use. I Am A Passionate Writer And I Aim To Make My Books Hold Information That Makes Them Worthwhile Reading. Thank You Again!

If You Would Be So Kind To Leave My Book A Positive Star Rating I Would Highly Appreciate That And I Will Continue Trying To Make My Books Make A Worthwhile Impact That Is Worth Your Investment As Well. Farewell.

Made in United States
North Haven, CT
23 April 2023